SECOND EDITION

Zoology

LABORATORY MANUAL

An Integrative Approach

D. BRENT BURT
MATTHEW A. KWIATKOWSKI
JUSTIN SULLIVAN

Kendall Hunt
publishing company

Cover images © Shutterstock, Inc.

Kendall Hunt
publishing company

www.kendallhunt.com
Send all inquiries to:
4050 Westmark Drive
Dubuque, IA 52004-1840

ISBN 978-1-4652-8301-6

Printed in the United States of America

Contents

Introduction to the Compound Light Microscope and Hypothesis Testing

LAB 1

This lab will introduce you to the use and care of the compound light microscope and other standard equipment used in an introductory biology lab. You will also be introduced to how the scientific method is used to investigate questions about the natural world.

Objectives

- Understand the parts and operation of a compound light microscope.
- List the steps for bringing an object into focus with the compound light microscope.
- Understand magnification.
- Understand processes and terminology associated with scientific methodology.

Introduction to the Compound Microscope

In a biology lab, the compound microscope is one of the most widely used instruments ever invented. Proper care and use of this precise piece of equipment will ensure that you gain a complete understanding of its function. It is important that all proper procedures be followed when using the compound microscope. Failure to do so can result in damage to the microscopes and/or an inability to complete lab exercises in a timely manner.

You will be assigned a microscope and microscope slide number that you will use throughout the semester. Daily checks will be made of both the microscopes and microscope slides to ensure that all procedures are being followed. If you suspect that your microscope is damaged or is not working properly, immediately notify your lab instructor. Never attempt repairs on your own!

Parts of the Compound Microscope

Using Figure 1.1, identify the parts of the microscope and understand their functions on your microscope.

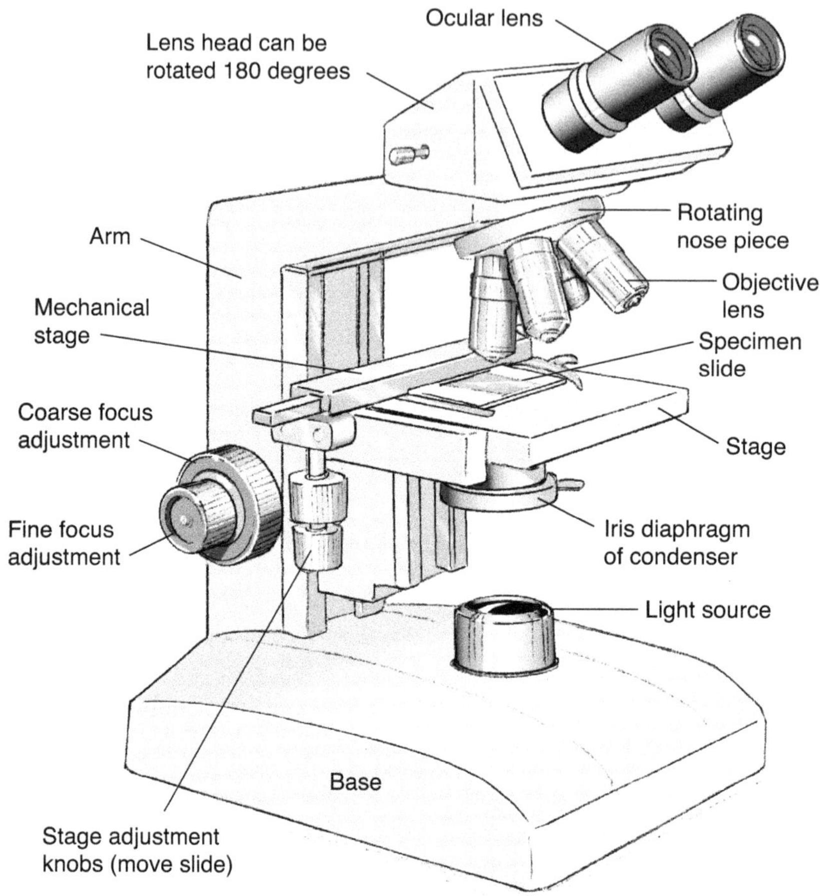

Figure 1.1 Compound Microscope.

Total Magnification

Magnification using a compound microscope is achieved through the interaction of two sets of lenses, the ocular lens and the objective lens (often microscopes have multiple objective lenses that can be rotated such that the desired magnification can be selected). To determine the **total magnification** of the object being viewed multiply the magnification of the ocular lens by the magnification of the objective being use. Complete **Table 1.1** to list the total magnifications for which you are responsible in this lab.

Table 1.1 Total Magnifications

Objective Lens Name	Objective Lens Magnification	Ocular lens magnification	Total Magnification
Scanning objective lens			
Low power objective lens			
Medium power objective lens			
High power objective lens			

Care and Proper Use of the Compound Microscope

Make sure that you have a clutter free work area to set you microscope (remove any bags or notebooks). When instructed, go to the microscope cabinet and obtain your microscope following the procedure listed below:

1. Grasp the arm of the microscope with one hand (your dominant hand) and support the base of your microscope with the opposite hand.
2. Place the microscope carefully on the table with the cord keeper facing away from you.
3. Place your microscope a few inches from the edge of the lab table to allow for comfortable viewing. **Do not drag the microscope across the table!**
4. Adjust your lab chair to a comfortable viewing height.
5. Loosen the set screw on the side of the microscope head and rotate the oculars so that they are facing you.
6. Starting with the scanning objective in place (**always start with the scanning objective in place first**), plug the microscope into the outlet and turn on the power switch.
7. Adjust the lighting to a comfortable intensity (use a combination of the iris diaphragm and the light intensity knob).
8. Inspect the microscope. If you notice any damage or if it is dirty or dusty, notify your lab instructor.
   a. If necessary, clean the ocular lenses with optical lens paper

Using the Microscope

Using a specimen designated by your lab coordinator, complete the steps below. **Make sure you follow the steps carefully as failure to do so could lead to broken or damaged equipment.**

1. Place the specimen on the slide and secure the slide to the microscope using the stage clip. Check that the slide is held by the stage clip by moving it with the mechanical stage control.
2. Rotate the scanning power (4X) objective into place on the nosepiece.
   a. When focusing, always start at low power and work your way up.
3. Using the mechanical stage control, position the specimen over the stage aperture.
4. Use the coarse adjustment to raise the stage to its highest position.
   a. Looking through the microscope oculars, turn the coarse adjustment slowly away from you, lowering the stage until your specimen comes into focus.
   b. Remember to adjust the iris diaphragm for the best image.
5. Make final adjustments to your image with the fine adjustment knob.
6. Center your specimen and rotate the low power (10X) objective into place.
   a. As your microscope is parfocal, only minor adjustments with the fine adjustment knob should be needed to bring the specimen back into focus.

7. Center your specimen and rotate medium power (20X) objective into place on the nosepiece.
   a. As before, use the fine adjustment knob to bring the specimen back into focus.
8. Center your specimen and rotate high power (40X) objective into place on the nosepiece.
   a. **Caution!** Make sure you only use the fine adjustment knob to make any necessary corrections. Failure to do so may result in equipment damage.

Microscope Exercises

Image Inversion

The mirrors in the microscope flip and reverse the image before it reaches your eyes. This is called image inversion. Using the steps outlined previously for getting an image in focus, complete the steps listed below.

1. Place the "Letter e" slide in the stage clips taking care to ensure the letter is right-side up, from your perspective.
2. Get the letter "e" in focus using the scanning power objective.
3. Briefly describe how that "letter e" appears when viewed through the microscope.

   ____

   ____

4. Looking through the microscope, use the mechanical stage controls to move your slide to the right. In which direction does the image move? ____________
5. Looking through the microscope, use the mechanical stage controls to move your slide forward (away from you). In which direction does the image move? ____________
6. Leave the "letter e" slide in your microscope as you will need if the next exercise.

Field of View

The field of view is everything visible when you look through the eyepiece. When you change the microscope from the scanning objective lens to the low or high power objective lens, the field of view will decrease, causing you to see less of the object or less of an area that you could see while using the scanning objective. Therefore, the scanning objective lens has the greatest field of view. Awareness of the field of view is important when viewing a slide. If the object is in focus when in the scanning objective lens, but the object is not in the center of the field of view, you may not see the object when you change to the next higher-power objective.

1. With the "e" already focused and in the center of the field of view, move the objective lens from scanning to the low power objective lens.
2. When you look through the oculars, the field of view will be smaller and the "e" may be partially out of focus. Use the fine adjustment knob to bring the "e" back into clear focus.
3. Once the "e" is in focus using the low power objective lens, draw the "e" as you see it in your field of view using the circle provided below.
4. After you have drawn the "e," move the objective lens from low power to the medium power objective lens.
5. When you look through the oculars, the field of view will be smaller and the "e" may be partially out of focus. Use the fine adjustment knob to bring the "e" back into clear focus.

6. Once the "e" is in focus using the medium power objective lens, draw the "e" as you see it in your field of view using the circle provided below.
7. After you have drawn the "e," move the objective lens from medium power to the high power objective lens. Although it may look like the high power objective lens might touch the slide, it will not, provided you had the "e" in clear focus on the low power objective lens.
8. When you look in the eyepiece, the field of view is even smaller and what you see of the "e" may be partially out of focus. Use the fine adjustment knob to bring the "e" into clear focus. If you cannot see any part of the "e" when you switched to the high power objective lens, switch back to the low power objective lens and make sure the "e" is in clear focus and in the center of the field of view.
9. Once you have any part of the "e" in clear focus, draw what you see in your field of view using the circle provided below.

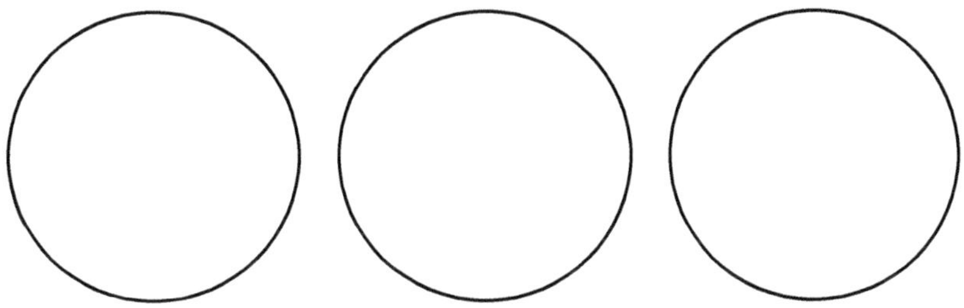

low power medium power high power

Depth of Field

Depth of field refers to the total thickness of an object when it is in clear focus. Depth of field varies with the different objectives and is inversely related to magnification. Complete the steps listed below.

1. Using the scanning power objective, examine a prepared slide of three colored threads mounted on top of each other.
2. Using the mechanical stage controls, move an area where all three threads are overlapping into the center of your field of view.
3. After bringing the image into clear focus, use the fine adjustment knob to adjust the stage up and down and try to determine the order of the threads from top to bottom.
   a. Note: The order of the threads will not be the same on all slides.
4. Complete steps 2 and 3 with the low power objective in place.
5. Complete steps 2 and 3 with the medium power objective in place.
6. Complete steps 2 and 3 with the high power objective in place.
7. Record your observations, relative to which color of thread is uppermost, middle, and lowest.
   a. Top thread color ____________________
   b. Middle thread color ____________________
   c. Bottom thread color ____________________
8. Did the order of the threads appear to change from one magnification to the next?
9. Which objective allowed you to have the clearest view of the thread order?

____

Introduction to the Scientific Method

The scientific method is a systematic approach to an observation or inquiry that allows you to study events occurring in the natural world. One assumption held by the scientific methods is that the events in question are both observable and measurable (i.e. cannot be explained on the basis of faith or spirituality). The scientific method requires you to follow a set of steps outlined below. Each step will be explained and discussed further during the lab period.

Steps of the Hyopothetico-Deductive Method

1. Make Observations of a Natural Phenomenon
   a. Can be made directly from nature or through other written works
2. Ask Questions
   a. Asking how or why the observed phenomenon exists.
3. State Hypothesis
   a. Tentative answer to your question
   b. Must be testable and falsifiable
   c. Set up two hypotheses:
      1. Define the null hypothesis (abbreviated H_o)
         - ______________________________________
      2. Define the alternative hypothesis (abbreviated H_a)
         - ______________________________________
   d. Use previous studies or observations to make predictions
      1. Predictions should be made using deductive reasoning and should offer specifics about the results of the experiment designed in the next step.
4. Collect Data to Test Hypotheses and Predictions
   a. Design study
      1. One key feature of a good experimental design is the prediction and control of variables that might influence the variables of interest in your experiment.
         - What is a variable?
           a. ______________________________________
         - Define an independent variable.
           a. ______________________________________
         - Define a dependent variable.
           a. ______________________________________
   b. Conduct study
      1. Collect data on how change in the independent variable affects the dependent variable in each test group.
5. Analyze data
   a. Data analysis varies depending on the design of the study but at a minimum will include some comparison of the mean, median, range and standard deviation.
      1. Define mean
         - ______________________________________
      2. Define median
         - ______________________________________

3. Define range
   - ____
4. Define standard deviation
   - ____

6. Make Conclusions and Publish Results
   b. Hypothesis tests have two possible outcomes:
      1. Reject the null hypothesis and conclude the alternative is supported.
      2. Fail to reject the null hypothesis and conclude that there is not enough evidence to suggest the null hypothesis is false.
   c. If rejected, rethink hypothesis and retry study

Hypothetic-Deductive Model Exercise

Obtain a peer-reviewed journal article from your lab instructor and complete the following:

1. What are the hypotheses? ____
   a. What is the null hypothesis?
      i. ____
   b. What is the alternative hypothesis?
      i. ____
2. What are the main variables being investigated?
   a. Identify the independent variable(s).
      i. ____
   b. Identify the dependent variable(s).
      i. ____
3. What are the main conclusions?
   a. ____
      ____
      ____

Foraging and Predator/ Prey Relationships

LAB
2

This lab investigates the foraging success of a predator on prey of different sizes. Larger prey might provide more energy to a predator, but they can also require more time to capture and ingest. In this lab, you will collect data on the capture and ingestion time for a predator, for prey of different sizes.

Objectives

- Form hypotheses regarding the foraging success of a predator.
- Present a predator with prey of different sizes.
- Collect and analyze data on predator foraging.
- Form conclusions based on collected data.

Before You Go to Lab

Complete Exercises I and II prior to coming to lab.

Exercises

I. Background Knowledge

Using outside sources approved by your instructor (e.g., your textbook), answer the following:

- To what phylum does your predator species belong? ____
- Does your predator species exhibit intra- or extracellular digestion? What do these terms mean?
- Does your predator species have a complete or an incomplete digestive system? What is the difference between these two types of digestive systems?
- How does your predator species capture prey—what structures are involved?

- What are some potential limitations or advantages to this method of prey capture?

- To what phylum does your prey species belong? ______________________________

II. Exploring the Scientific Literature and Hypotheses

Conduct a search on Google Scholar using the terms "prey size" and "prey size foraging success." Read through the abstracts of some of the scientific articles that are listed. What do the studies suggest is a general relationship between prey size and predator foraging?

Hypotheses and Predictions

Based on the scientific articles you found, come up with the following hypotheses and predictions regarding prey capture:

- What is a null hypothesis regarding capture time and ingestion time of the different-sized prey?

- What is the alternative hypothesis regarding capture time and ingestion time of the different-sized prey?

- Based on the literature you read for this exercise, make predictions about the capture time and ingestion time for the different prey sizes.

III. Procedures

Materials and Methods

In this study your methods will involve introducing two differently sized prey items to a predator and recording capture and ingestion times. Your instructor will provide a list of materials and instructions for testing your hypotheses on the effects of prey size on predator capture and ingestion times.

Figure 2.1. *Acheata domesticus* and *Anolis carolinensis*. These are examples of prey and predators that may be used in your study. Other pairs that could be used include *Hydra* sp. and *Daphnia* sp., *Daphnia* sp. and *Danio* sp., etc.

IV. Results

Record your results in Table 2.1.

Table 2.1

	Small Prey		*Large Prey*	
Group Number	**Capture Time**	**Ingestion Time**	**Capture Time**	**Ingestion Time**

Calculate the average capture time and ingestion time for each prey size.

	Small Prey		*Large Prey*	
	Capture Time	Ingestion Time	Capture Time	Ingestion Time
Average time:				

V. Conclusions

Complete a lab report that contains the following:

- An introduction including both your null and alternative hypotheses, and a purpose statement.
- A description of your materials and methods, including statistical methods.
- A results section discussing your initial findings
- A copy of your data in table format and any figures used to interpret your data.
- A conclusion citing two scholarly articles which answers the following questions:

1. Did your results support your null hypothesis?

2. Were there differences in capture time for the different prey sizes?

3. Were there differences in ingestion time for the different prey sizes?

4. How do your group's results compare to those of the entire class?

Save the document as "Lab 2 Report (Last names of group members)" and submit it to the dropbox as instructed.

LAB 3

Cardiovascular

The cardiovascular system in animals circulates materials throughout the body. These materials may include respiratory gases, nutrients, wastes, and parts of the immune system. The rate of circulation can vary among animals based on metabolism. This lab investigates the effects of temperature on the cardiovascular physiology in an ectotherm.

Objectives

- Form hypotheses regarding the effects of temperature on heart rate in an ectotherm.
- Test these hypotheses using a *Daphnia* model.
- Collect and analyze data on *Daphnia* heart rate.
- Form conclusions based on collected data.

Before You Go to Lab

Complete Exercises I and II prior to coming to lab.

Exercises

I. Background Knowledge

Using outside sources approved by your instructor (e.g., your textbook), answer the following:

- The organisms you'll be working with, *Daphnia*, are ectotherms. What is an ectotherm?

- What is the difference between ectotherms and endotherms when it comes to physiological processes?

- Are most animals endotherms or ectotherms? ______________________________

- Does *Daphnia* have a closed or open circulatory system? What is the difference between these two types of circulatory systems?

II. Exploring the Scientific Literature and Hypotheses

Conduct a search on Google Scholar using the term "ectotherm heart rate." Read through the abstracts of some of the scientific articles that are listed. What are some variables that influence this physiological process?

Hypotheses and Predictions

Based on the scientific articles you found, come up with the following hypotheses and predictions regarding temperature and heart rate in *Daphnia*:

- What is a null hypothesis regarding the association between temperature and heart rate in *Daphnia*?

- What is the alternative hypothesis regarding the association between temperature and heart rate in *Daphnia*?

- Based on the literature you read for this exercise, make predictions about the effects of different temperatures on heart rate in *Daphnia*.

- Do you predict there will be any differences in the effects of temperature on heart rate between *Daphnia* that are different sizes? Why or why not?

III. Procedures

Materials

Daphnia magna
D. pulex
Microscope
Depression slides
Bulb pipettes
Stopwatch

****NOTE:** Keep the light source on your scope **off** except when directly viewing the *Daphnia*.

Before you start, determine the temperature of the three water treatments in which the *Daphnia* are being kept. Record those temperatures both below and in Table 3.1 in the Results section.

Daphnia magna

Cold Treatment: __________°C

Room Temperature: __________°C

Warm Treatment: __________°C

Daphnia pulex

Cold Treatment: __________°C

Room Temperature: __________°C

Warm Treatment: __________°C

Working in pairs, each member of the lab group should complete the following procedures.

Procedure A

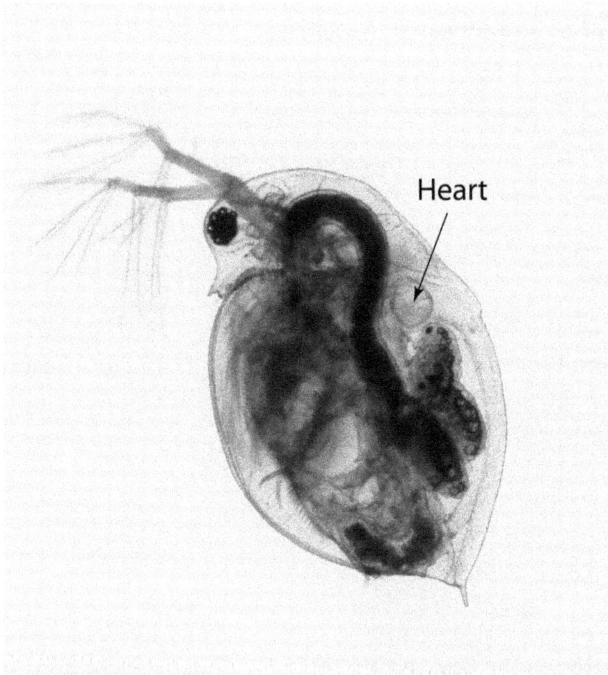

Figure 3.1. Daphnia with dorsal heart indicated by arrow.

It is important to work quickly during this procedure. (Why?)

1. Obtain a depression slide.

2. Using a bulb pipette, transfer a *D. pulex* (with water) from the treatment container into the depression on the slide. Try to keep the drop of water small so the *Daphnia* does not swim out of your field of view.

3. Observe the *D. pulex* under low power on your microscope. Find the heart on the *Daphnia* (use Figure 1 as a reference).

4. With a lab partner using a stopwatch to time you, record the *Daphnia*'s heart rate for **15 seconds**.

5. Record the number of heartbeats in Table 1.

6. Fill in the heartbeats recorded for your group members as well.

7. Rinse the *Daphnia* into the designated container.

8. Repeat steps 1–7 for the other two temperature treatments.

Procedure B

1. Repeat steps 1–8 from Procedure A, but this time using *D. magna*. Record your results in Table 2.

IV. Results

Procedure A

Record the number of heartbeats for *D. pulex* in Table 3.1 under the columns labeled "15 sec." Once you are done with your observations, calculate the number of beats per minute (BPM) for each observation and record that number in the "BPM" columns.

Table 3.1

	Cold Treatment ______°C		*Room Temperature* ______°C		*Warm Treatment* ______°C	
Student Name	**15 sec**	**BPM**	**15 sec**	**BPM**	**15 sec**	**BPM**

Using a spreadsheet software program, record the data for the entire class. Calculate the mean heart rate for *D. pulex* in beats per minute for each temperature treatment.

	Cold Treatment ______°C	*Room Temperature* ______°C	*Warm Treatment* ______°C
Mean BPM:			

Procedure B

Record the number of heartbeats for *D. magna* in Table 3.2 under the columns labeled "15 sec." Once you are done with your observations, calculate the number of beats per minute (BPM) for each observation and record that number in the "BPM" columns.

Table 3.2

	Cold Treatment ______°C		*Room Temperature* ______°C		*Warm Treatment* ______°C	
Student Name	15 sec	BPM	15 sec	BPM	15 sec	BPM

Using a spreadsheet software program, record the data for the entire class. Calculate the mean heart rate for *D. magna* in beats per minute for each temperature treatment.

	Cold Treatment ______°C	*Room Temperature* ______°C	*Warm Treatment* ______°C
Mean BPM:			

Relationship between Heart Rate and Temperature

Using a spreadsheet software program, plot the data for temperature and mean heart rate as directed by your Instructor. Determine the relationship between heart rate and temperature by calculating the slope of the line from (1) cold temperature to room temperature, and (2) from room temperature to warm temperature.

Slope = $\Delta y / \Delta x = y_2 - y_1 / x_2 - x_1$

Enter the values for the slopes of *D. pulex*:

Cold temperature/room temperature slope: ________

Room temperature/warm temperature slope: ________

Enter the values for the slopes of *D. magna*:

Cold temperature/room temperature slope: ________

Room temperature/warm temperature slope: ________

V. Conclusions

Complete a lab report that contains the following:

- An introduction including both your null and alternative hypotheses, and a purpose statement.

- A description of your materials and methods, including statistical methods.

- A results section discussing your initial findings

- A copy of your data in table format and any figures used to interpret your data.

- A conclusion citing two scholarly articles which answers the following questions:

  1. Did the results support your null hypothesis?

  2. Were there differences in heart rate among the three temperatures?

  3. For every degree of increase in temperature, how much should heart rate change?

  4. Were there differences in the effects of temperature between the different sizes of *Daphnia*?

     - Explain why you think there were or were not.

Save the document as “Lab 3 Report (last names of group members)” and submit it to the dropbox as instructed.

LAB 4

Muscular System

Muscle cells allow movement through contraction. These contractions are the result of numerous chemical reactions between proteins within each muscle cell. The cumulative force of an entire muscle can be influenced by variables such as temperature and body size. This lab will investigate the physiology of skeletal muscle using arthropods as a model system.

Objectives

- Form hypotheses regarding the effects of body size and temperature on muscle activity.
- Test these hypotheses using arthropods.
- Collect and interpret data on muscle performance.
- Form conclusions based on collected data.

Before You Go to Lab

Complete Exercises I and II prior to coming to lab.

Exercises

I. Background Knowledge

Using outside sources approved by your instructor (e.g., your textbook), answer the following:

- The organisms you'll be working with are arthropods. What is an arthropod?

- What is the difference between arthropods and vertebrates in terms of how skeletal muscles are connected to the skeletal system?

- The organisms you'll be working with are ectotherms. What effect might temperature have on muscle contraction? Would you expect differences between endotherms and ectotherms in muscle contractions? Explain.

II. Exploring the Scientific Literature and Hypotheses

Temperature and Muscle Performance

In this lab you will investigate the effects of temperature on jump distance in crickets, *Acheta domesticus.* Conduct a search on Google Scholar using the term "cricket temperature jumping." Read through the abstracts of some of the scientific articles that are listed to answer the following questions (you may use your own terms for an additional search to find more articles).

- What is a null hypothesis regarding the association between temperature and the jump distance of the crickets?

- What is the alternative hypothesis regarding the association between temperature and the jump distance of the crickets?

- Based on the literature you read for this exercise, make predictions about the effects of different temperatures on the jumping distance of crickets.

Muscle Force and Body Size

In this lab you will also investigate the maximal force generated by arthropods that differ in size. Conduct a search on Google Scholar using the term "insect body size muscle force." Read through the abstracts of some of the scientific articles that are listed to answer the following questions (you may use your own terms for an additional search to find more articles).

- What is a null hypothesis regarding the association between muscle force and body size?

- What is the alternative hypothesis regarding the association between muscle force and body size?

- Based on the literature you read for this exercise, make predictions about the effects of body size on muscle force. Do you predict the larger arthropod will produce greater force than the smaller one? What if you control for body size? What does it mean to "control for body size"?

III. Procedures

Temperature and Muscle Performance

Materials

Crickets *(A. domesticus)*
Measuring tape
Holding container

Before you start, record the temperatures at which the crickets have been kept. Record those temperatures **both below and in Table 4.1 in the Results section.**

Cold Treatment: __________°C

Room Temperature: __________°C

Warm Treatment: __________°C

Work together as a group for the following procedures.

Procedure A

It is important to work quickly during this procedure. (Why?)

1. Obtain a cricket from one of the temperature treatments and place it in the designated jumping area on the start line.

2. Gently prod the cricket until it jumps. Have your group members carefully observe the point from which the cricket jumped and the point where it lands **after a single jump**. Capture your cricket and place a marker at the point where the jump began and where it landed.

3. Carefully measure the length of the cricket as well as the distance of the jump. Record the values in Table 4.1. Place the cricket in the designated container.

4. Repeat steps 1–3 with 4 more crickets from the same temperature treatment. Record the results of each in Table 4.1.

5. Repeats steps 1–4 with 5 crickets from each of the other two temperature treatments (for a total of 15 crickets).

Body Size and Muscle Force

Materials

Large arthropod specimen
Small arthropod specimen
Weights
Digital balance
Holding/Weighing container

Procedure B

1. Your instructor will provide your group with two arthropods of different sizes and a "sled" to attach.

2. Attach the "sled" as instructed to your arthropod.

3. Beginning with a small amount of weight, see if the arthropod can pull the weight.

4. Add weight until the arthropod cannot pull the load.

5. Record these weights in Table 4.2.

6. With your instructor's help, carefully detach the "sled" from the arthropod.

7. Place the arthropod in its holding container.

8. Place an empty holding container on the digital balance and tare (zero) the balance. Replace the empty holding container with the one containing your arthropod. Record the mass of the arthropod in the correct column of Table 4.2.

9. Record the data from the other groups in class in Table 4.2.

IV. Results

Temperature and Muscle Performance

Procedure A

Record the jump distance and body length of your crickets in Table 4.1. Once you are done with you observations, calculate the mean jump distance and record it below.

Table 4.1

	Cold Treatment ______°C			*Room Temperature* ______°C			*Warm Treatment* ______°C		
Cricket	**Jump Distance (mm)**	**Body Length (mm)**	**Jump Distance/ Body Length**	**Jump Distance (mm)**	**Body Length (mm)**	**Jump Distance/ Body Length**	**Jump Distance (mm)**	**Body Length (mm)**	**Jump Distance/ Body Length**
1									
2									
3									
4									
5									

	Cold Treatment ______°C	*Room Temperature* ______°C	*Warm Treatment* ______°C
Mean Jump Distance/Body Size:			

Relationship between Temperature and Jump Distance/Body Size Ratio

Using a spreadsheet software program, plot the data for temperature and jump distance/body size ratio. Determine the relationship between temperature and jump distance/body size ratio by calculating the slope of the line from (1) cold temperature to room temperature, and (2) from room temperature to warm temperature.

Enter the values for the slopes:

Cold temperature/room temperature slope: _________

Room temperature/warm temperature slope: __________

Body Size and Muscle Force

Table 4.2

	Large arthropod			*Small arthropod*		
Group	Weight Pulled	Arthropod Mass	Weight Pulled/ Arthropod Mass	Weight Pulled	Arthropod Mass	Weight Pulled/ Arthropod Mass
1						
2						
3						
4						
5						
6						
7						
8						
9						
10						
11						
12						
13						
14						

	Large arthropod	Small arthropod
Mean Weight Pulled/Body Mass:		

V. Conclusions

Complete a lab report that contains the following:

Temperature and Muscle Performance

- An introduction including both your null and alternative hypotheses, and a purpose statement.
- A description of your materials and methods, including statistical methods.
  - Why did we calculate the jump distance/body size ratio and use that as a measure of performance rather than just jump distance?
- A results section discussing your initial findings

- A copy of your data in table format and any figures used to interpret your data.
- A conclusion citing two scholarly articles which answers the following questions:

1. Did the results support your null hypothesis?
2. Why did we control for body length?
3. Were there differences in jumping distance among the three temperatures?
4. For every degree of increase in temperature, how much should jump distance change?
5. Would you expect this relationship to continue if you looked at different temperatures? Why or why not?

Body Size and Muscle Force

- An introduction including both your null and alternative hypotheses, and a purpose statement.
- A description of your materials and methods, including statistical methods.
- A results section discussing your initial findings
- A copy of your data in table format and any figures used to interpret your data.
- A conclusion citing two scholarly articles which answers the following questions:

1. Did the results support your null hypothesis?
2. When you controlled for body size (mass), did the larger arthropods generate more muscle force than the smaller arthropods?
3. What other variables might influence muscle force?

Save the document as "Lab 4 Report (last names of group members)" and submit it to the dropbox as instructed.

LAB 5

Phylogenetics

In this lab we investigate how extensive phenotypic and genetic changes can evolve in organisms over a large timescale. These major changes are due to the long-term accumulation of microevolution changes (due to natural selection, mutation, and genetic drift) in lineages that become reproductively isolated and form new branches in the evolutionary tree of animal life (i.e., the animal phylogeny). Understanding the distribution of major phenotypic traits among organisms allows biologists to discover the branching patterns in the phylogeny of animals.

Objectives

- Explain how reproductive isolation leads to the formation of new lineages or branches in an evolutionary tree or phylogeny.
- Explain how genetic and phenotypic changes accumulating independently in each branch lead to divergences in the gene pools and morphological/behavioral traits associated with each lineage.
- Use trait variation among a group of organisms to reconstruct the best estimate of their phylogenetic relationships.
- Explain how different classes of traits lead to the formation of monophyletic, paraphyletic, or polyphyletic groupings on a phylogenetic tree.

Before You Go to Lab

Be sure you read over the terms listed below and understand their usage in phylogenetic biology.

- Autapomorphy—a trait that has changed in a lineage at the tip of a phylogeny. This trait indicates how this lineage is unique, differing from all other lineages.
- Clade—a grouping of lineages in a phylogeny that includes an ancestral node and all of its descendants.
- Homology—trait similarity among organisms due to common ancestry.
- Homoplasy—trait similarity among organisms that is not due to common ancestry. The traits may be similar due to adapting to similar environmental challenges in similar ways.
- Monophyletic—another term for a clade.
- Node—a point in a phylogenetic tree where a lineage splits into two lineages, which may then begin to diverge genetically, morphologically, and behaviorally.
- Paraphyletic—a grouping of lineages in a phylogeny that includes an ancestral node and some, but not all, of its descendants.
- Phylogeny—an evolutionary tree showing the pattern of relatedness among lineages.
- Polyphyletic—a grouping of lineages in a phylogeny that do not share a common ancestral node.
- Symplesiomorphy—trait similarity derived from an ancestral lineage that is shared among some, but not all, of the descendants of this common ancestor. Another term for a shared ancestral homology.
- Synapomorphy—a trait that is modified in the most recent common ancestral lineage of a clade and is retained in all of the descendant lineages. Another term for a shared derived homology.

Exercises

Ia.

Simulate the evolution of a group of lizard species using the "Evolving Lizards" exercise in the program Evolutionary Evidence of the SimBio Virtual Labs package.

- Click on the SimBio Virtual Labs icon on the computer desktop.
- Click on the Evolutionary Evidence link.
- Close the Getting Started popup window.
- Select the Evolving Lizards option in the dropdown menu at the top of the screen.

You now have a notebook view showing five islands, one of which is occupied by lizards. One the right side of the notebook is a list of morphological features that might be found on a lizard. You are now ready to simulate the evolution of lizards as they migrate to each of the islands and evolve different morphological features when they become reproductively isolated.

- Click on a lizard on the occupied island. Notice that it lacks any of the possible traits.
- Drag one of these lizards to another island. Double click on the lizard on the new island and give it one or two of the traits listed.

You have just simulated the migration of lizards to this island and the evolution of new traits for this population via mutation.

- Run the simulation for 60–120 years before stopping it.

This simulates the establishment of many generations of populations of lizards with the chosen traits.

- Continue to drag a lizard from your most recently established (last) island to an unoccupied island, give it one or more of the traits, and run the simulation for 60–120 generations, repeating the process until all of the islands are populated with lizards with different trait combinations.
- Be sure to write down the order of the migration events (the order of islands populated) and the traits given to each population in Table 5.1.

Table 5.1 Simulation Step History

Step	Island #	Trait Newly Evolved
Start (original island)		---------------------
2		
3		
4		
5		

Ib.

You have simulated the movement of lizards from island to island and the establishment of populations that have evolved novel traits in reproductive isolation. Now let's see if you can use the traits from another lab group's simulation to reconstruct that group's pattern of species evolution.

- Switch computers with another group but leave your simulation results on your computer.
- Drag a lizard from each population onto the notebook
- Place lizards with similar traits next to each other and line up all the lizards in a single row at the top of the notebook (Figure 5.1).

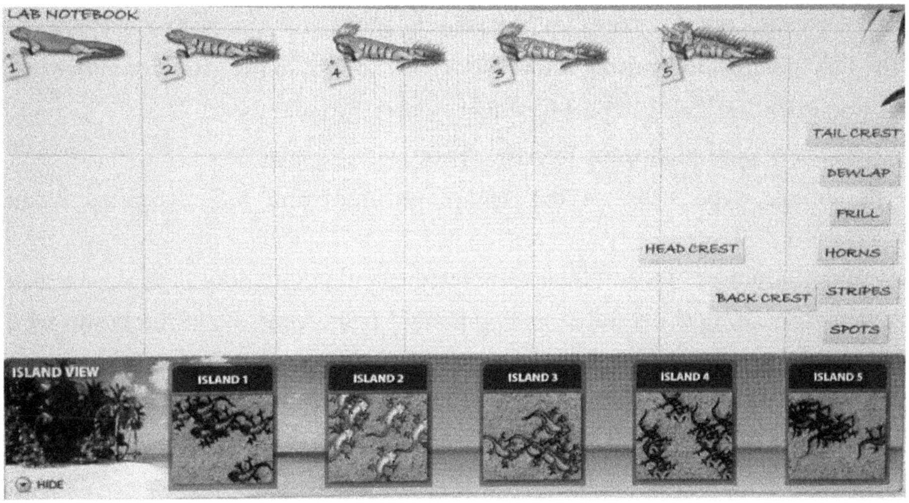

Figure 5.1. A row of lizards from each island ordered by trait similarity.

- **Without looking at the original group's simulation step history**, draw your best estimate of the phylogeny of these lizards using the pencil tool. Hide the island view to make more room if needed.
- Next, indicate the pattern of evolution of each trait by dragging the trait labels from the right side of the notebook to the appropriate branches of the tree (Figure 5.2).

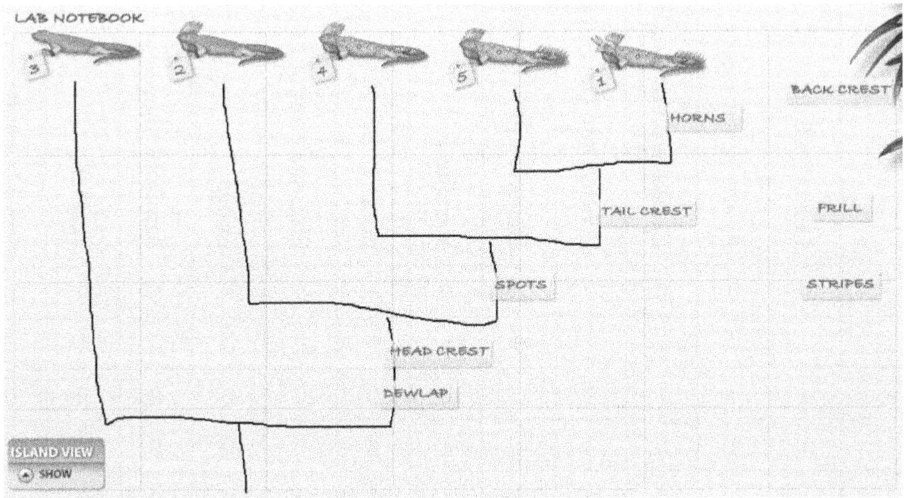

Figure 5.2. Phylogeny of lizards with appropriate placement of associated traits.

The traits placed on the internal branches of the tree represent synapomorphies (Figure 5.2). These traits indicate that all the descendants of this branch have inherited this trait from a common ancestor. Synapomorphies are evidence that these groups of descendants represent monophyletic groups or clades.

Figure 5.2 indicates that having a tail crest supports monophyly of lizards on islands 1 and 5. A tail crest and spots indicate that lizards from islands 1 4, and 5 represent a more inclusive clade. A dewlap and head crest are a synapomorphies for an even larger clade including species 1, 2, 4, and 5. The remaining trait (horns) represent an autapomorphy, or a trait that indicates how species 1 is unique, but it does not help us understand how different lizard populations are related.

- Using the copy tool, save a screenshot of your phylogeny with the proper placement of the traits.
- Open the Lab 5 Report form using Microsoft Word and fill in the group member names.
- Save the document as "Lab 5 Report (last names of group members)."
- Paste the image of your phylogeny into the report.
- Below the image, type a list of the synapomorphies and the associated clades they support (ex. "Clade 1,2,4").
- Also list the autapomorphies and their associated island population(s).
- Insert a table indicating the simulation step history from someone in the group who conducted this simulation.
- Compare your resulting phylogeny with the simulation step history conducted by the original group. If your phylogeny differs from the simulation step history, discuss these differences.
- Save the file and keep it open because you will be adding to it soon.

II.

Examine trait variation among the eight animals below and reconstruct their phylogenetic history.

- Use the following data matrix to guide you in the construction of the phylogeny. A 0 indicates the absence of a trait, while a 1 indicates the trait is present in that animal.

	Cephalization	Bilateral Symmetry	Exoskeleton of Chitin	Protostome Development	Deuterostome Development	Vertebrae	Jaws	Endoskeleton of Bone	Lungs	Amniotic Eggs	Feathers	Toothless Bill	Wings	Long Boney Tail
Sponge	0	0	0	0	0	0	0	0	0	0	0	0	0	0
Shrimp	1	1	1	1	0	0	0	0	0	0	0	0	0	0
Dragonfly	1	1	1	1	0	0	0	0	0	0	0	0	1	0
Lamprey	1	1	0	0	1	1	0	0	0	0	0	0	0	1
Bass	1	1	0	0	1	1	1	1	0	0	0	0	0	1
Sparrow	1	1	0	0	1	1	1	1	1	1	1	1	1	0
Eagle	1	1	0	0	1	1	1	1	1	1	1	1	1	0
Crocodile	1	1	0	0	1	1	1	1	1	1	0	0	0	1

- Use character labels to indicate synapomorphies on the appropriate branches that support your phylogenetic conclusions.

In Exercise I, all of the lizard traits presented to you represented either synapomorphies or autapomorphies. The data matrix in this exercise contains synapomorphies for the most part, but also includes an example of a shared ancestral homology (symplesiomorphy) and a homoplasy.

- Draw the phylogeny for these animals (with the appropriately labeled synapomorphies) at the end of your Lab 5 Report.
- Identify and document which trait is a symplesiomorphy and which is a homoplasy in your Lab 5 Report. These will be the traits that don't fit on the tree with a single evolutionary change of the character. That is, these traits will require extra evolutionary "steps".
- Your final task is to briefly state why we use synapomorphies to indicate evolutionary relationships and not symplesiomorphies or homoplasies. Consider the data matrix used here. How do we determine which trait is a synapomorphy and which is not? **Hint:** Look up the term *parsimony*.
- Save the finalized lab report as "Lab 5 Report (last names of group members).", and turn the document in as instructed.

LAB 6

Cnidarians

Cnidarians are a group demonstrating how animals with relatively simple body plans can display behavioral and morphological adaptations to meet the variety of environmental challenges they face.

Objectives

- Demonstrate knowledge of structures and associated functions seen in cnidarians.
- Demonstrate knowledge of and the ability to work in groups to draw a phylogenetic tree with proper placement of key synapomorphic traits for lineages covered in this lab and other closely related taxa.
- Demonstrate the ability to read and critically assess an article from the primary literature both in writing and orally during group discussions.

Before You Go to Lab

1. Complete Exercise I prior to coming to lab. Completing this section will prepare you for the remaining lab exercises and help prepare you for the pre-lab quiz.
2. Attempt Exercise III with your own estimate of the group's phylogeny and associated character placement so that members of your group will be able to discuss differing views and complete the assignment during lab.
3. Read the primary literature article linked to this lab. Download and fill out a Journal Article Review Form to document your critical assessment of this paper. Save the file as "Lab 6 JAR (your last name)." Turn the completed document in as instructed.

Exercises

I. Form and Function

Briefly summarize how the following structures help cnidarians complete the physiological challenges in the following six categories.

Key Terms and Structures

- Cnidocytes
- Diploblastic
- Extracellular digestion
- Incomplete digestive tract
- Medusa
- Nematocyst
- Nerve net
- Planula larva
- Polyp
- Radial symmetry

1. Support and Movement

2. Foraging and Digestion

3. Gas Exchange and Circulation

4. Water Balance and Excretion

5. Development

6. Nervous and Sensory Systems

II. Investigations of Specimens

A. Obtain a prepared slide of a *Hydra* longitudinal section. Identify and understand the function of each structure and label figure 6.1.

- Cnidocil
- Cnidocyte
- Epidermis
- Gastrodermis
- Mesogloea
- Gastrovascular cavity
- Mouth
- Nematocyst
- Tentacles

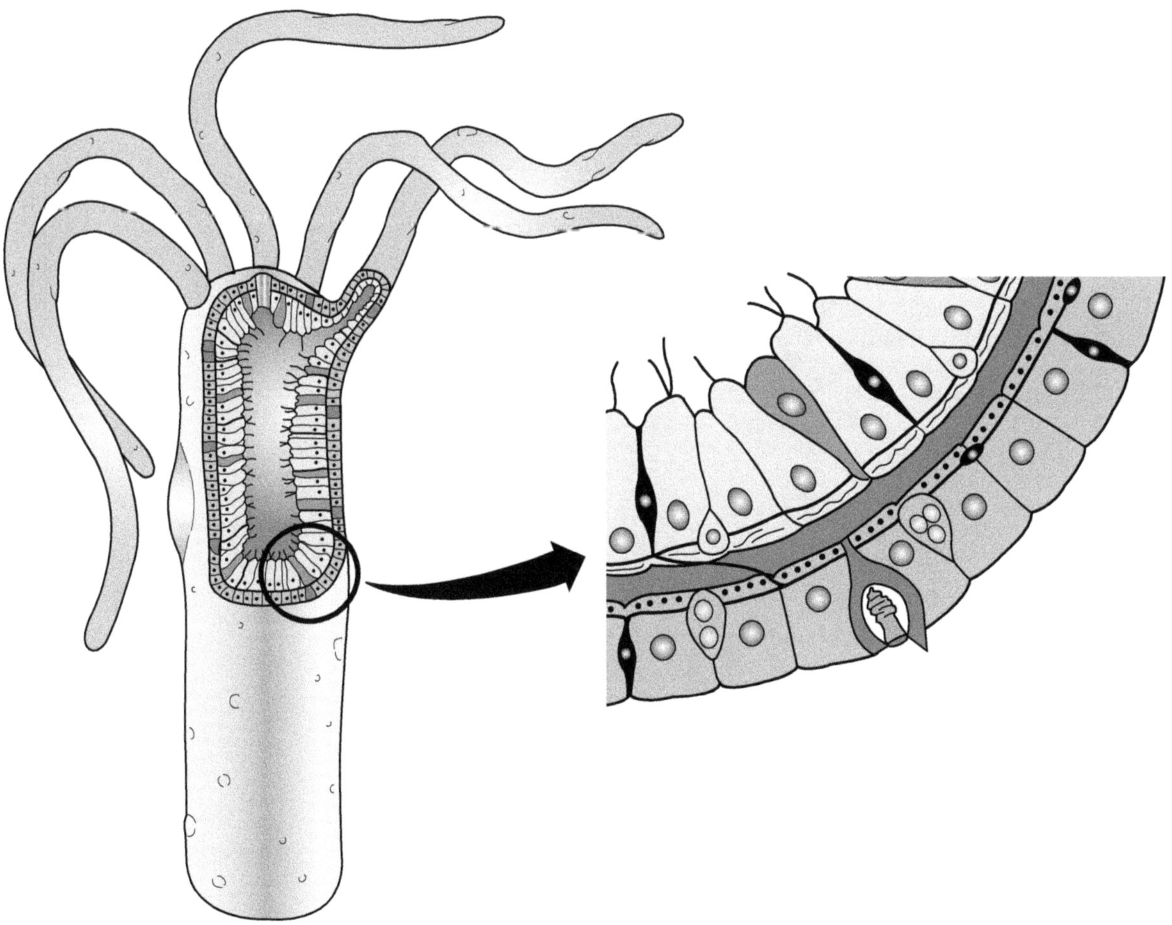

Figure 6.1. *Hydra.*

B. Obtain a medusa body form of *Gonionemus*. Identify and understand the function of each structure and label figure 6.2.

- Aboral/exumbrella surface
- Gastrovascular cavity
- Gonads
- Manubrium
- Mesoglea
- Mouth
- Oral lobe
- Oral/subumbrella surface
- Radial canal
- Ring canal
- Tentacles
- Velum

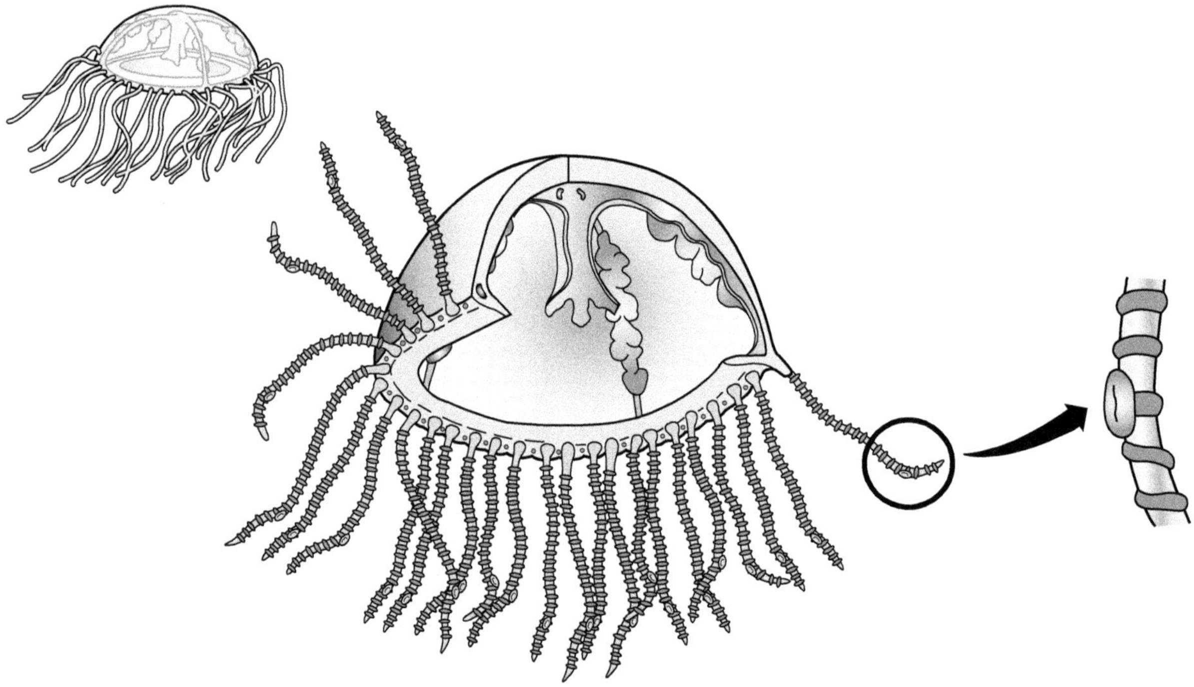

Figure 6.2. *Gonionemus.*

C. Obtain slides of an *Obelia* colony and *Obelia* medusa. Identify and understand the function of each structure and label figure 6.3.

Colony:

- Coenosarc
- Gastrovascular cavity
- Gonangium
- Gonotheca
- Hydranth
- Hydrotheca
- Perisarc

Medusa:

- Manubrium
- Mouth
- Radial canal
- Ring canal
- Tentacles

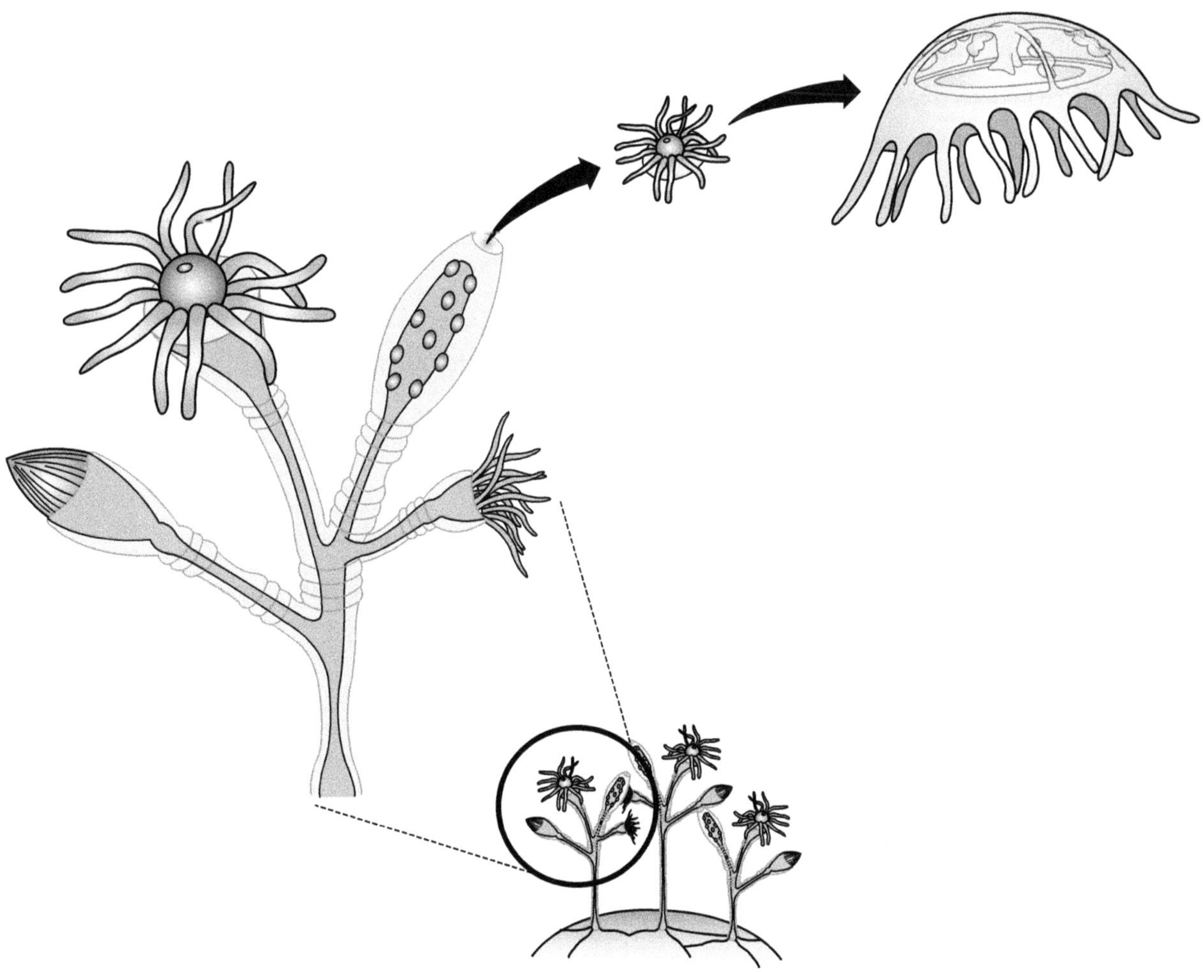

Figure 6.3. *Obelia* colony and *Obelia* medusa.

III. Phylogenetic Placement

Work in your group to draw a phylogeny that includes the following lineages: Choanoflagellates, Porifera, Anthozoa, Scyphozoa, Cubozoa, Hydrozoa, Ctenophora. Map the following traits on the interior branches of the tree:

- Biradial symmetry
- Boxlike medusa body form
- Cnidocytes
- Collagen
- Colloblasts
- Complete digestive tract
- Eukaryotic cell form
- Extracellular digestion
- Gemmule
- Gut with septal filaments
- Incomplete digestive tract
- Internal system of pores and canals
- Intracellular digestion
- Medusa body form
- Mouth surrounded by tentacles
- Multicellular, incipient tissues
- Nerve net system
- Planula larvae
- Polyp reduced or lost
- Radial polyp body form
- Radial symmetry
- Rhopalium
- Siphonoglyph
- Solitary or simple colony of single-celled organisms
- Spicules
- Spongin
- Strobilation
- Velarium
- Velum

Use the lab report template provided by your Instructor, save the file as "Lab 6 Report (last names of group members)," and turn the document in as instructed.

IV. Journal Article Review

Critically review the article provided by the Instructor and answer the following questions:

1. What was the purpose of the study? What questions were the researcher(s) asking?

2. What methods were utilized?

3. What were the main conclusions?

4. What evidence supported those conclusions?

5. Why were those conclusions important?

V. Examples of the Representative Taxa

Make notes about the various examples of each taxa on display.

Lophotrochozoa I–Flatworms and Rotifers

LAB 7

This lab focuses on two lineages in the Lophotrochozoan clade: The phyla Platyhelminthes and Rotifera. Animals in these two lineages lack both of the characteristics for which the clade is named (the lophophore respiratory/feeding structure and trochophore larvae) but nonetheless share a common ancestor with other lineages in this clade.

Objectives

- Demonstrate knowledge of body types, structures, and associated functions seen in rotifers and flatworms.
- Demonstrate knowledge of and the ability to work in groups to draw a phylogenetic tree with proper placement of key synapomorphic traits for lineages covered in this lab and the previous lab.
- Demonstrate the ability to work in groups to form a hypothesis, collect data, produce figures from the data, and use the figures to form conclusion.
- Demonstrate the ability to communicate effectively all aspects of your group's study.
- Demonstrate the ability to assess critically your study and another group's study.

Before You Go to Lab

1. Complete Exercise I prior to coming to lab. Completing this section will prepare you for the remaining lab exercises and help prepare you for the pre-lab quiz.
2. Attempt Exercise III with your own estimate of the group's phylogeny and associated character placement so that members of your group will be able to discuss differing views and complete the assignment during lab.
3. Plan for Exercise IV. Formulate a hypothesis and design a study to collect data that you could analyze to test this hypothesis. This preparation will allow you to compare different study options with your group members once you come to lab.

Exercises

I. Form and Function

Briefly summarize how the following structures help the animals covered in this lab complete the physiological challenges in the following six categories. How do these traits compare to those in lineages covered in the previous lab?

Key Terms and Structures

- Acoelomate
- Adhesive toe
- Bilateral symmetry
- Central nervous system
- Cephalization
- Complete digestive tract
- Corona
- Cutaneous respiration
- Eyespots
- Free-living/parasitic
- Hydrostatic skeleton
- Incomplete digestive tract
- Mastax
- Pedal glands
- Proglottid
- Protonephridium
- Protostome
- Pseudocoelomate
- Scolex
- Triploblastic

1. Support and Movement

2. Foraging and Digestion

3. Gas Exchange and Circulation

4. Water Balance and Excretion

5. Development

6. Nervous and Sensory Systems

II. Investigations of Specimens

A. Identify and understand the function of the following structures in a prepared slide of a rotifer and label figure 7.1.

- Corona
- Pedal glands
- Foot
- Toe
- Mastax
- Eyespots

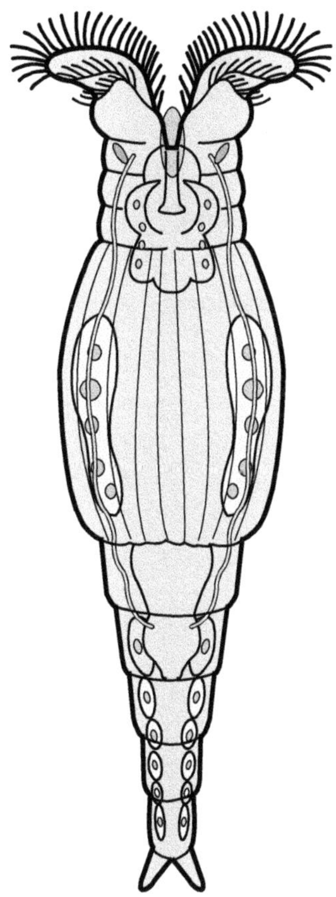

Figure 7.1. Rotifer.

Obtain living rotifers and observe them under a microscope to gain a more complete appreciation of the animals' behaviors.

B. Identify and understand the function of the following structures in a prepared slide of a turbellarian flatworm and label figure 7.2.

- Eyespots
- Pharynx
- Intestine

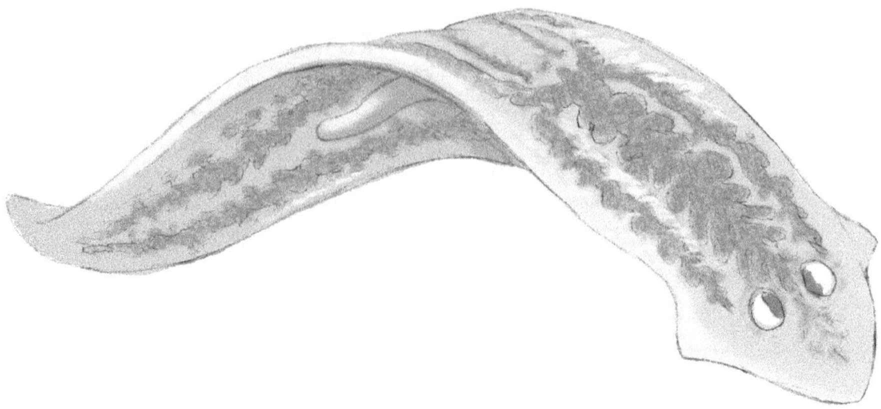

Figure 7.2. Turbellarian flatworm.

C. Identify and understand the function of the following structures in a prepared slide of a cestode flatworm and label figure 7.3.

- Scolex
- Proglottid
- Genital pore
- Sperm duct
- Testes
- Vagina
- Ovaries
- Uterus
- Mature eggs
- Mehlis's gland
- Vitelline gland
- Immature proglottid
- Mature proglottid
- Gravid proglottid

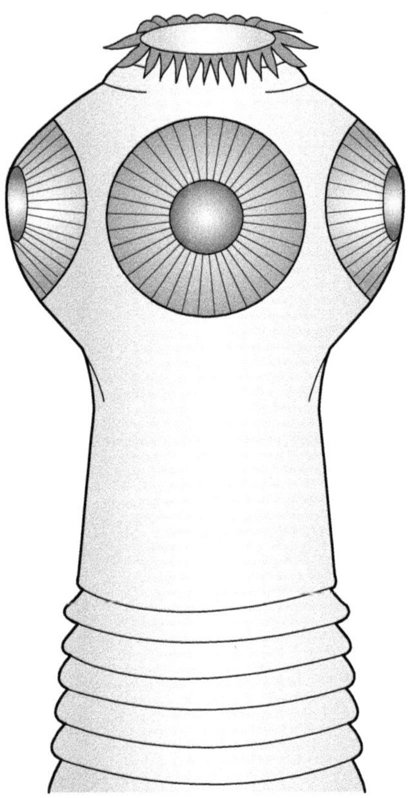

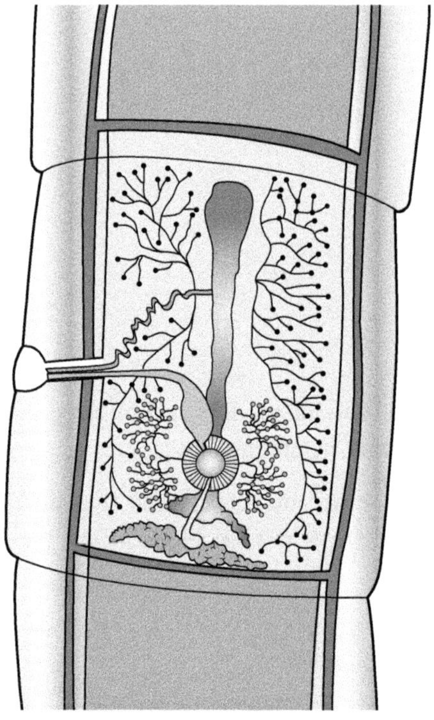

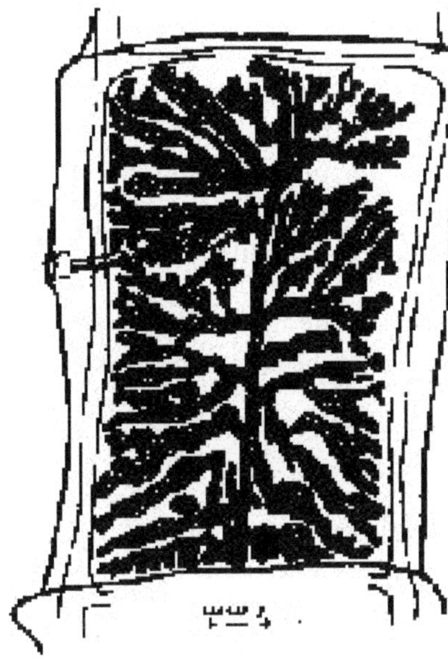

Figure 7.3. Cestode.

III. Phylogenetic Placement

Work in your group to draw a phylogeny that includes the lineages from the last lab plus the Rotifera, Turbellaria, Trematoda, and Cestoda. Map the following traits on the interior branches of the tree:

- Protostome development
- Bilateral symmetry
- Cephalization
- Central nervous system
- Triploblasty
- Body cavity
- Protonephridium
- Rhabdites
- Anterior adhesive organ
- Posterior adhesive organ
- Mehlis's gland
- Neodermis
- Branched intestine

Use the lab report template provided by your Instructor to complete this exercise, save the document as "Lab 7 Phylogeny (last names of group members)," and turn the document in as instructed.

IV. Mini Study

Work in your group to collect behavioral data on flatworms.

A. Turbellarian flatworms demonstrate phototaxis.

- Form a hypothesis as to whether these flatworms demonstrate negative or positive phototropism and then design and conduct a study to test your hypothesis.
- Enter your data into an Excel worksheet and produce appropriate figures that will allow you to draw conclusions.

- Complete a lab report that contains the following:
  - An introduction including both your null and alternative hypotheses, and a purpose statement.
  - A description of your materials and methods, including statistical methods.
  - A results section discussing your initial findings
  - A copy of your data in table format and any figures used to interpret your data.
  - A conclusion section citing two scholarly articles.
  - A critique of your study: How would you do things differently in the future?

You will likely not have time in lab to complete the analyses and report. Discuss when and where ALL group members will meet to complete the analyses of the data and writing of the report.

- Save the document as "Lab 7 Mini Study (last names of group members)" and turn the document in as instructed.

V. Examples of the Representative Taxa

Make notes about the various examples of each taxa on display.

Lophotrochozoa II–Molluscs and Annelids

LAB
8

This lab focuses on the two lineages in the Lophotrochozoan clade with the most species: the phyla Mollusca and Annelida.

Objectives

- Demonstrate knowledge of body types, structures, and associated functions seen in molluscs and annelid worms.
- Demonstrate knowledge of and the ability to work in groups to draw a phylogenetic tree with proper placement of key synapomorphic traits for lineages covered in this lab and the previous lab.
- Demonstrate the ability to read and critically assess an article from the primary literature both in writing and orally during group discussions.

Before You Go to Lab

1. Complete Exercise I prior to coming to lab. Completing this section will prepare you for the remaining lab exercises and help prepare you for the pre-lab quiz.
2. Review instructions for details on techniques needed to properly dissect your specimens.
3. Attempt Exercise III with your own estimate of the group's phylogeny and associated character placement so that members of your group will be able to discuss differing views and complete the assignment during lab.
4. Read the primary literature article linked to this lab. Download and fill out a Journal Article Review Form to document your critical assessment of this paper. Save the file as "Lab 8 JAR (your last name)." Turn the completed document in as instructed.

Exercises

I. Form and Function

Briefly summarize how the following structures help the animals covered in this lab complete the physiological challenges in the following six categories. How do these traits compare to those in lineages covered in the previous lab?

Key Terms and Structures

- Bilateral symmetry
- Central nervous system
- Cephalization
- Coelom
- Complete digestive tract
- Cutaneous respiration
- External shell
- Foot
- Free-living/parasitic
- Gills
- Hydrostatic skeleton
- Lungs
- Metanephridium
- Open/closed circulation
- Parapodia
- Protostome
- Radula
- Setae
- Triploblastic
- Trochophore

1. Support and Movement

2. Foraging and Digestion

3. Gas Exchange and Circulation

4. Water Balance and Excretion

5. Development

6. Nervous and Sensory Systems

II. Investigations of Specimens

A. Review instructions for details on techniques needed to properly dissect your specimen and find the following structures. Identify and understand the function of each structure and label figure 8.1.

- Anterior adductor
- Excurrent siphon
- Foot
- Gills
- Gonad
- Heart (ventricle and atrium)
- Incurrent siphon
- Intestine
- Labial palp
- Mantle
- Mouth
- Nephridium
- Pericardium
- Posterior adductor
- Stomach and digestive gland
- Umbo
- Valve

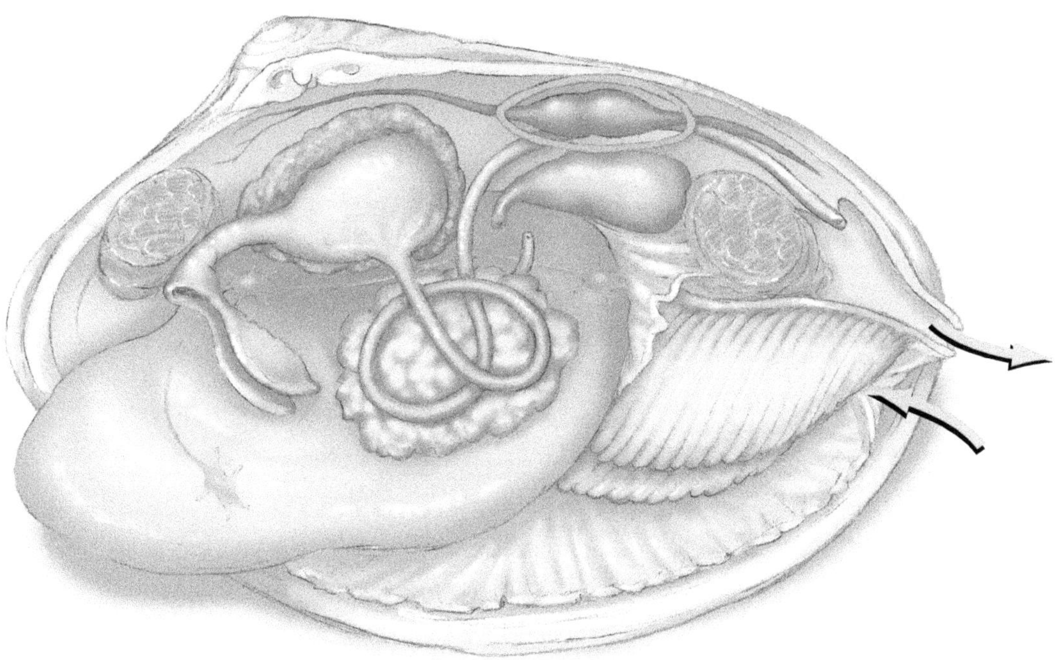

Figure 8.1. Mussel.

B. Review instructions for details on techniques needed to properly dissect your specimen and find the following structures. Identify and understand the function of each structure and label figure 8.2.

- Aortic arches
- Brain
- Clitellum
- Crop
- Dorsal vessel
- Egg funnel and oviduct
- Esophagus
- Gizzard
- Intestine
- Mouth
- Nerve cord
- Ovary
- Pharynx
- Seminal receptacle
- Seminal vesicles
- Ventral vessel

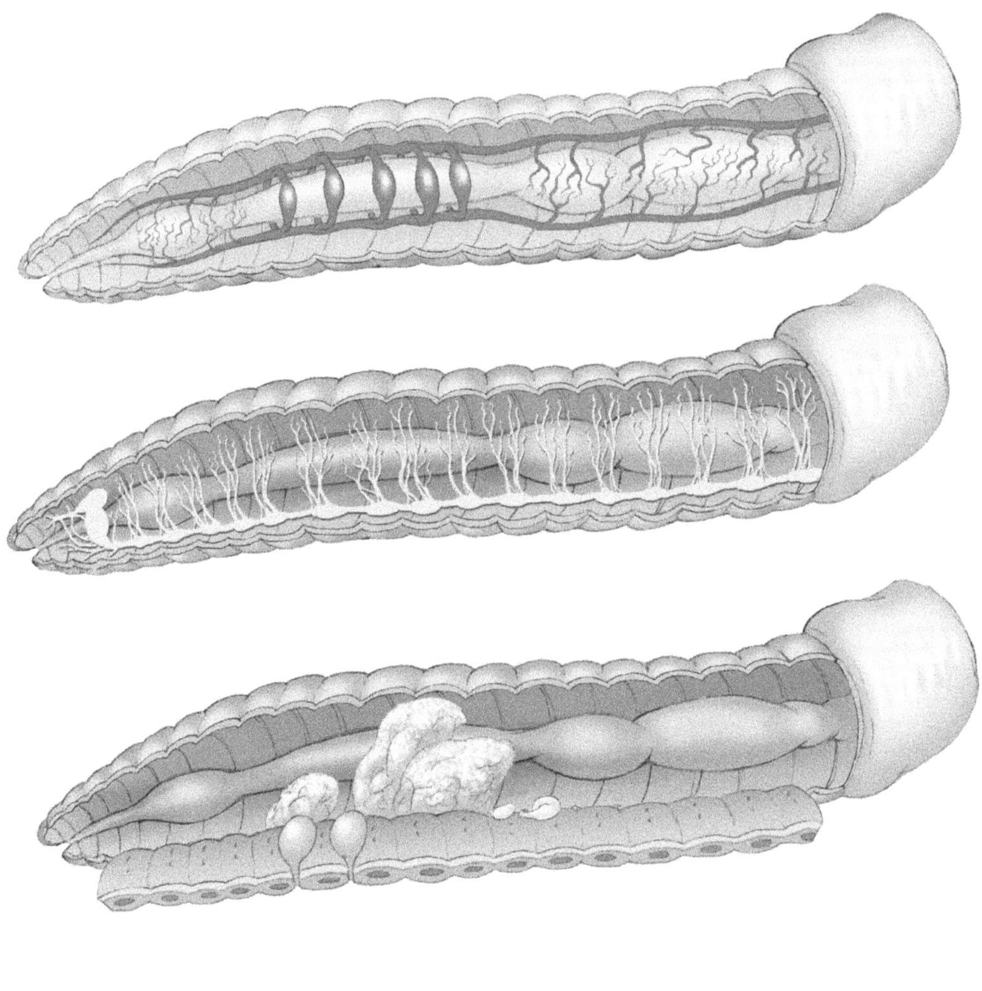

Figure 8.2. Annelid.

Understand the function of the following structures and identify them in a prepared slide of an earthworm cross section. Label figure 8.3.

- Intestinal lumen
- Typhlosole
- Dorsal vessel
- Ventral vessel
- Nerve cord
- Nephridium
- Setae
- Circular muscle
- Longitudinal muscle
- Epidermis
- Cuticle

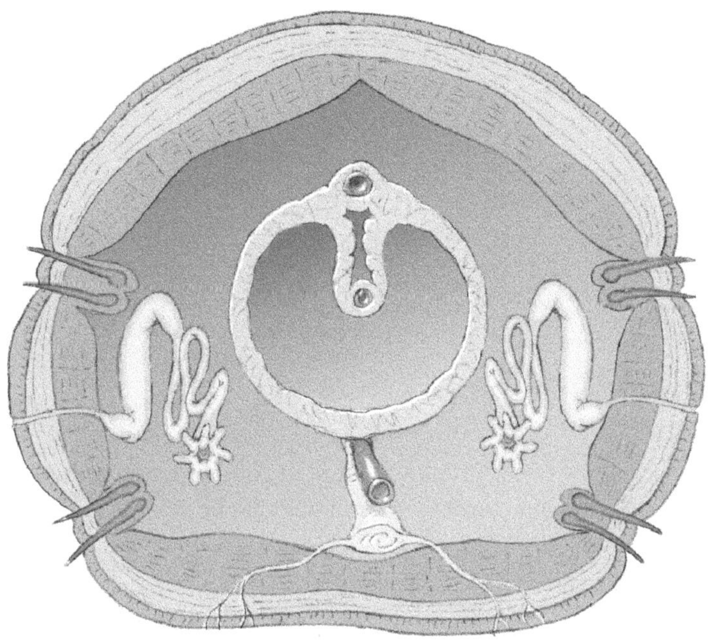

Figure 8.3. Earthworm cross section.

III. Phylogenetic Placement

Work in your group to draw a phylogeny that includes the lineages from the last lab plus the Polyplacophora, Gastropoda, Cephalopoda, Bivalvia, Scaphopoda, Errantia, and Sedentaria. Map the following traits on the interior branches of the tree:

- Radula
- Muscular foot
- Mantle
- Shell with multiple plates
- Single shelled
- Two-valved shell
- Gills
- Lungs
- Torsion
- Well-developed head
- Siphuncle
- Captacula
- Metameric body plan
- Parapoda
- Trochophore larva
- Clitellum
- Setae

Use the lab report template provided by your Instructor to complete this exercise, save the document as "Lab 8 Phylogeny (last names of group members)," and turn the document in as instructed.

IV. Journal Article Review

Critically review the article provided by the Instructor and answer the following questions:

1. What was the purpose of the study? What questions were the researcher(s) asking?

2. What methods were utilized?

3. What were the main conclusions?

4. What evidence supported those conclusions?

5. Why were those conclusions important?

V. Examples of the Representative Taxa

Make notes about the various examples of each taxa on display.

LAB 9

Ecdysozoa I-Nematodes, Arthropoda I—Chelicerates

This lab focuses on the two lineages in the Ecdysozoa clade. The phylum Nematoda, commonly referred to as the roundworms, contains a diverse group representing possibly half a million free-living and parasitic species. The phylum Arthropoda represents and even larger clade of animals, and in fact contains more species than all other groups put together. We will therefore continue our study of arthropods in Lab 10.

Objectives

- Demonstrate knowledge of body types, structures, and associated functions seen in roundworms and chelicerate arthropods.
- Demonstrate knowledge of and the ability to work in groups to draw a phylogenetic tree with proper placement of key synapomorphic traits for lineages covered in this lab and the previous lab.
- Demonstrate the ability to work in groups to form a hypothesis, collect data, produce figures from the data, and use the figures to form conclusions.
- Demonstrate the ability to communicate effectively all aspects of your group's study.
- Demonstrate the ability to assess critically your study and another group's study.

Before You Go to Lab

1. Complete Exercise I prior to coming to lab. Completing this section will prepare you for the remaining lab exercises and help prepare you for the pre-lab quiz.
2. Attempt Exercise III with your own estimate of the group's phylogeny and associated character placement so that members of your group will be able to discuss differing views and complete the assignment during lab.
3. Read the primary literature article linked to this lab. Download and fill out a Journal Article Review Form to document your critical assessment of this paper. Save the file as "Lab 9 JAR (your last name)." Turn the completed document in as instructed.

Exercises

I. Form and Function

Briefly summarize how the following structures help the animals covered in this lab complete the physiological challenges in the following six categories. How do these traits compare to those in lineages covered in the previous lab?

Key Terms and Structures

- Bilateral symmetry
- Book gills
- Book lungs
- Central nervous system
- Cephalization
- Chelicerae
- Coelom
- Complete digestive tract
- Compound eyes
- Cutaneous excretion
- Cutaneous respiration
- Cuticle
- Exoskeleton
- Free-living/parasitic
- Gills
- Hydrostatic skeleton
- Malpighian tubules
- Maxillary/antennal glands
- Open circulation
- Protostome
- Tagmata
- Tracheae
- Triploblastic

1. Support and Movement

2. Foraging and Digestion

3. Gas Exchange and Circulation

4. Water Balance and Excretion

5. Development

6. Nervous and Sensory Systems

II. Investigations of Specimens

A. Understand the function of the following structures and identify them in a prepared slide of a roundworm cross section. Label figures 9.1A and 9.1B.

- Cuticle
- Dorsal nerve cord
- Excretory duct
- Ovary
- Oviduct
- Uterus
- Testes
- Hypodermis
- Intestine
- Muscle
- Pseudocoel

Figure 9.1A. Roundworm (Female) cross section.

Figure 9.1B. Roundworm (Male) cross section.

B. Identify and understand the function of the following external structures in a horseshoe crab. Label figure 9.2.

- Abdomen
- Anus
- Book gills
- Carapace
- Chelicera
- Chelicera
- Chilarium
- Compound eye
- Genital operculum
- Gill operculum
- Gnathobase
- Mouth
- Pedipalp
- Simple eye
- Telson

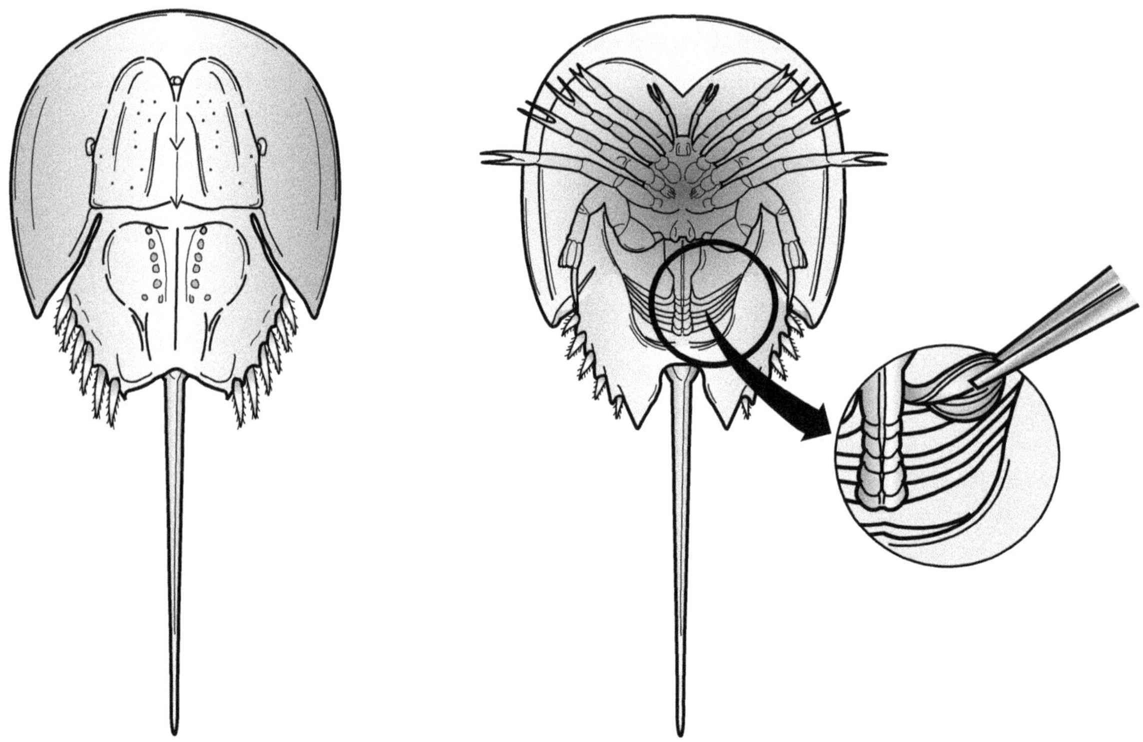

Figure 9.2. Horseshoe crab.

C. Identify and understand the function of the following external structures in a spider. Label figure 9.3.

- Abdomen
- Cephalothorax
- Walking legs
- Eyes
- Chelicerae
- Pedipalp

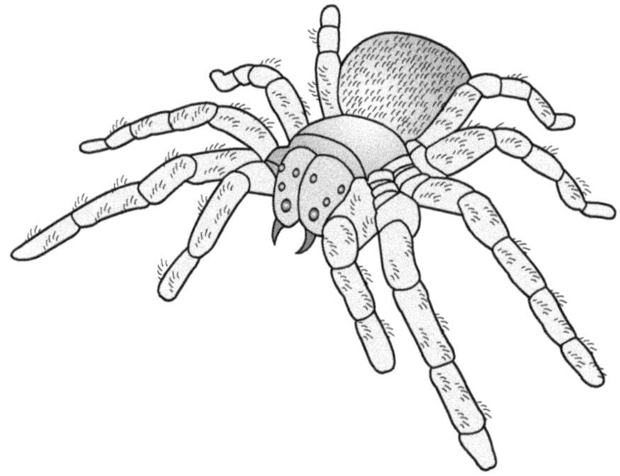

Figure 9.3. Spider.

III. Phylogenetic Placement

Work in your group to draw a phylogeny that includes the lineages from the last lab plus the Nematoda, Pycnogonida, Merostomata, Diplopoda, Chilopoda and Arachnida. Map the following traits on the interior branches of the tree:

- Cuticle
- Chelicarae
- Pedipalps
- Four pair of walking legs
- Ovigers
- Book gills
- Compound eye
- Shield-like cephalothorax
- First pair of legs modified as venom claws
- 2 pairs of legs per body segment

Use the lab report template provided by your Instructor to complete this exercise, save the document as "Lab 9 Phylogeny (last names of group members)," and turn the document in as instructed.

IV. Journal Article Review

Critically review the article provided by the Instructor and answer the following questions:

1. What was the purpose of the study? What questions were the researcher(s) asking?
2. What methods were utilized?
3. What were the main conclusions?
4. What evidence supported those conclusion?
5. Why were those conclusions important?

V. Examples of the Representative Taxa

Make notes about the various examples of each taxa on display.

Ecdysozoa II-Arthropoda II–Crustaceans and Insects

LAB 10

This lab focuses on the two additional arthropod lineages. The subphyla Crustacea and Hexapoda are lineages representing both large numbers of species and huge contributors to the animal biomass in virtually all ecosystems on the planet. Crustaceans are particularly well represented in aquatic ecosystems, with few terrestrial representatives (e.g., the isopods, or pillbugs). Insects are the main representatives of hexapods. They dominate the terrestrial environment and also have strong representation in freshwater habitats in juvenile stages of life.

Objectives

- Demonstrate knowledge of body types, structures, and associated functions seen in crustacean and hexapod arthropods.
- Demonstrate knowledge and the ability to work in groups to draw a phylogenetic tree with proper placement of key synapomorphic traits for lineages covered in this lab and the previous lab.
- Demonstrate the ability to work in groups to identify arthropod specimens to order using a dichotomous key.

Before You Go to Lab

1. Complete Exercise I prior to coming to lab. Completing this section will prepare you for the remaining lab exercises and help prepare you for the pre-lab quiz.
2. Review instructions for details on techniques needed to properly dissect your specimen.
3. Attempt Exercise III with your own estimate of the group's phylogeny and associated character placement so that members of your group will be able to discuss differing views and complete the assignment during lab.

Exercises

I. Form and Function

Briefly summarize how the following structures help the animals covered in this lab complete the physiological challenges in the following six categories. How do these traits compare to those in lineages covered in the previous lab?

Key Terms and Structures

- Antennae
- Antennal/maxillary/green gland
- Bilateral symmetry
- Biramous appendages
- Central nervous system
- Cephalization
- Coelom
- Complete digestive tract
- Compound eyes
- Cutaneous respiration
- Exoskeleton
- Gills
- Hemimetabolous metamorphosis
- Holometabolous metamorphosis
- Malphigian tubules
- Maxilla
- Nauplius larvae
- Open circulation
- Protostome
- Tagmata
- Tracheae
- Triploblastic
- Uniramous appendages
- Wings

1. Support and Movement

2. Foraging and Digestion

3. Gas Exchange and Circulation

4. Water Balance and Excretion

5. Development

6. Nervous and Sensory Systems

II. Investigations of Specimens

A. Identify and understand the function of the following external structures in a crayfish. Label figure 10.1.

- Abdomen
- Antenna
- Antennules
- Anus
- Cephalothorax
- Cheliped
- Maxillipeds
- Mouth
- Sternum
- Swimmerets
- Telson
- Tergum
- Uropod
- Walking legs

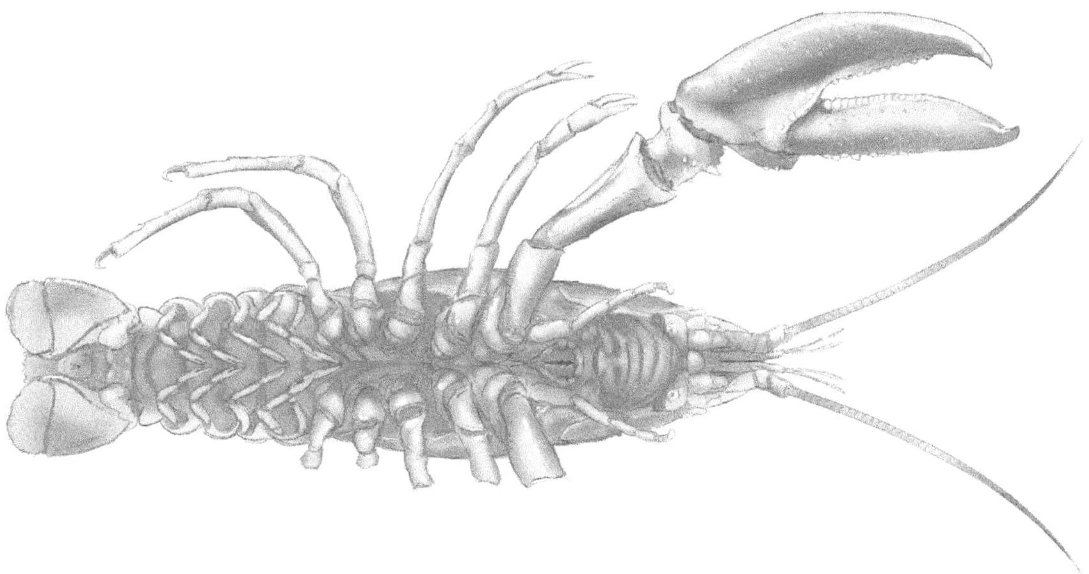

Figure 10.1 Crayfish.

B. Review instructions for details on techniques needed to properly dissect your specimen and find the following structures. Identify and understand the function of each structure and label figure 10.2.

- Antennal gland
- Stomach
- Brain
- Ventral nerve cord
- Digestive gland
- Heart
- Pericardium
- Ostium
- Gonads
- Intestine
- Gills (indicate the location on the figure 10.2)

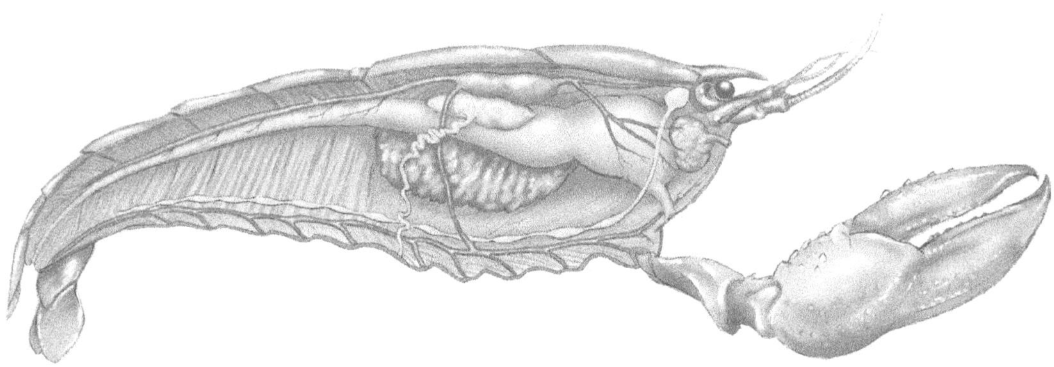

Figure 10.2 Crayfish.

III. Phylogenetic Placement

Work in your group to draw a phylogeny that includes the lineages from the last lab plus the Crustacea and Hexapoda. Map the following traits on the interior branches of the tree:

- Six legs
- Direct development
- Tracheal system
- Malpighian tubules
- Nauplius larvae
- Biramous second antennae
- Tripartite brain
- Mandibles

Use the lab report template provided by your Instructor, save the file as "Lab 10 Report (last names of group members).", and turn the document in as instructed.

IV. Using Dichotomous Keys

Dichotomous keys are important tools to help identify specimens in highly diverse groups such as arthropods. Work in your group using the provided dichotomous keys to identify the arthropods on your lab table to the correct order. Using the form provided by your Instructor, list the step history and identification for each specimen provided. More detailed keys exist to help identify specimens to more specific taxonomic units such as families, genera, and even species. Discuss some of the likely characters that would be included in these more detailed keys. Add a statement summarizing your discussion at the end of your key exercise report. Save the file as "Lab 10 Dichotomous Key (last names of group members", and turn it in as instructed.

V. Examples of the Representative Taxa

Make notes about the various examples of each taxa on display.

Deuterostomes I-Echinoderms

LAB 11

We began our survey of animal diversity with radially symmetrical animals that in large part were sessile. From this simple beginning, subsequent lineages evolved combinations of traits, including bilateral symmetry and cephalization, which afforded them more active lifestyles. This lab focuses on a major animal lineage that has abandoned bilateral symmetry and cephalization in the adult form but still maintains in many cases an active, yet slow-moving, lifestyle.

Objectives

- Demonstrate knowledge of body types, structures, and associated functions seen in echinoderms.
- Demonstrate knowledge of and the ability to work in groups to draw a phylogenetic tree with proper placement of key synapomorphic traits for lineages covered in this lab and the previous lab.
- Demonstrate the ability to read and critically assess an article from the primary literature both in writing and orally during group discussions.

Before You Go to Lab

1. Complete Exercise I prior to coming to lab. Completing this section will prepare you for the remaining lab exercises and help prepare you for the pre-lab quiz.
2. Review instructions for details on techniques needed to properly dissect your specimens.
3. Attempt Exercise III with your own estimate of the group's phylogeny and associated character placement so that members of your group will be able to discuss differing views and complete the assignment during lab.
4. Read the primary literature article linked to this lab. Download and fill out a Journal Article Review Form to document your critical assessment of this paper. Save the file as "Lab 11 JAR(your last name)." Turn the completed document in as instructed.

Exercises

I. Form and Function

Briefly summarize how the following structures help the animals covered in this lab complete the physiological challenges in the following six categories. How do these traits compare to those in lineages covered in the previous lab?

Key Terms and Structures

- Ambulacral ridge
- Ampulla
- Cardiac stomach
- Coelom
- Complete digestive tract
- Deuterostome
- Eye spot
- Gonads
- Lateral canal
- Madreporite
- Ossicles
- Papula
- Pedicellariae
- Podium
- Pyloric cecum
- Pyloric duct
- Pyloric stomach
- Radial canal
- Radial/pentaradial symmetry
- Respiratory tree
- Ring canal
- Spine
- Stone canal
- Sucker
- Water vascular system

1. Support and Movement

2. Foraging and Digestion

3. Gas Exchange and Circulation

4. Water Balance and Excretion

5. Development

6. Nervous and Sensory Systems

II. Investigations of Specimens

A. Review instructions for details on techniques needed to properly dissect your specimen and find the following structures. Identify and understand the function of each structure and label figure 11.1.

- Ambulacral ridge
- Ampulla
- Anus
- Cardiac stomach
- Coelomic cavity
- Eye spot
- Gonads
- Lateral canal
- Madreporite
- Ossicle
- Papula
- Podium
- Pyloric cecum
- Pyloric duct
- Pyloric stomach
- Radial canal
- Ring canal
- Spine
- Stone canal
- Sucker

Figure 11.1. Sea Star.

III. Phylogenetic Placement

Work in your group to draw a phylogeny that includes the lineages from the last lab plus the Crinoidea, Asteroidea, Ophiuroidea, Echinoidea, Holothuroidea, Chaetognatha, and Hemichordata. Map the following traits on the interior branches of the tree:

- Ambulacral grooves (open and closed)
- Arms broadly connected to central disc
- Deuterostome development
- Dorsal nerve cord
- Elongation of oral-aboral axis
- Endoskeleton of ossicles
- Fusion of ossicles to form a test
- Highly flexible arms
- Madreporite
- Pentaradial symmetry
- Pharyngeal gill slits
- Respiratory tree
- Suckers on tube feet
- Water vascular system

Use the lab report template provided by your Instructor, save the file as "Lab 11 Report (last names of group members).", and turn the document in as instructed.

IV. Journal Article Review

Critically review the article provided by the Instructor and answer the following questions:

1. What was the purpose of the study? What questions were the researcher(s) asking?

2. What methods were utilized?

3. What were the main conclusions?

4. What evidence supported those conclusions?

5. Why were those conclusions important?

V. Examples of the Representative Taxa

Make notes about the various examples of each taxa on display.

LAB 12

Vertebrata I–Fishes

This is the first of two labs examining vertebrates. In this lab we examine the most diverse group of vertebrates, the paraphyletic group we call "fish." This group is paraphyletic because tetrapods are derived from within one of the main fish lineages (the sarcopterygian, lobe-finned fishes).

Objectives

- Demonstrate knowledge of body types, structures, and associated functions seen in jawless, cartilaginous, ray-finned, and lobe-finned fishes.
- Demonstrate knowledge of and the ability to work in groups to draw a phylogenetic tree with proper placement of key synapomorphic traits for lineages covered in this lab and the previous lab.
- Demonstrate the ability to work in groups to form a hypothesis, collect data, produce figures from the data, and use the figures to form conclusions.
- Demonstrate the ability to communicate effectively all aspects of your group's study.
- Demonstrate the ability to assess your study critically.

Before You Go to Lab

1. Complete Exercise I prior to coming to lab. Completing this section will prepare you for the remaining lab exercises and help prepare you for the pre-lab clicker quiz.
2. Watch the *Fish Dissection* video for details on techniques needed to properly dissect your specimen. The video can be found in D2L.
3. Attempt Exercise III with your own estimate of the group's phylogeny and associated character placement so that members of your group will be able to discuss differing views and complete the assignment during lab.
4. Plan for Exercise IV. Do some basic research on shoaling behavior in fish to help you formulate a hypothesis, design a study, collect data, and then analyze the data to test your hypothesis. This preparation will allow you to compare different study options with your group members once you come to lab.

Exercises

I. Form and Function

Using the following list of key terms and structures, briefly summarize how animals in each of these groups are similar to and different from each other and how they compare to lineages covered in previous labs.

Key Terms and Structures:

- Ammocoete larvae
- Ampullae of Lorenzini
- Bilateral symmetry
- Bone
- Cartilage
- Central nervous system
- Cephalization
- Closed circulation
- Coelom
- Complete digestive tract
- Deuterostome
- Diphycercal tail
- Dorsal nerve cord
- Endostyle
- Eyes
- Gills
- Heterocercal tail
- Homocercal tail
- Jaws
- Kidney
- Lateral line
- Lobe fin
- Lungs
- Nostrils
- Notochord
- Operculum
- Paired appendages
- Pharyngeal slits
- Post-anal tail
- Ray fin
- Scales
- Swim bladder
- Triploblastic
- Vertebrae

1. Support and Movement

2. Foraging and Digestion

3. Gas Exchange and Circulation

4. Water Balance and Excretion

5. Development

6. Nervous and Sensory Systems

II. Investigations of Specimens

A. Review instructions for details on techniques needed to properly dissect your specimen and find the following structures. Identify and understand the function of each structure and label figure 12.1.

- Anal fin
- Anus
- Caudal fin
- First and second dorsal fins
- Gills
- Gonads (testes or ovaries)
- Heart
- Intestines
- Kidney
- Myomeres
- Pelvic and pectoral fins
- Pyloric ceca
- Spinal cord
- Spleen
- Stomach
- Urinary bladder
- Urogenital opening
- Vertebra

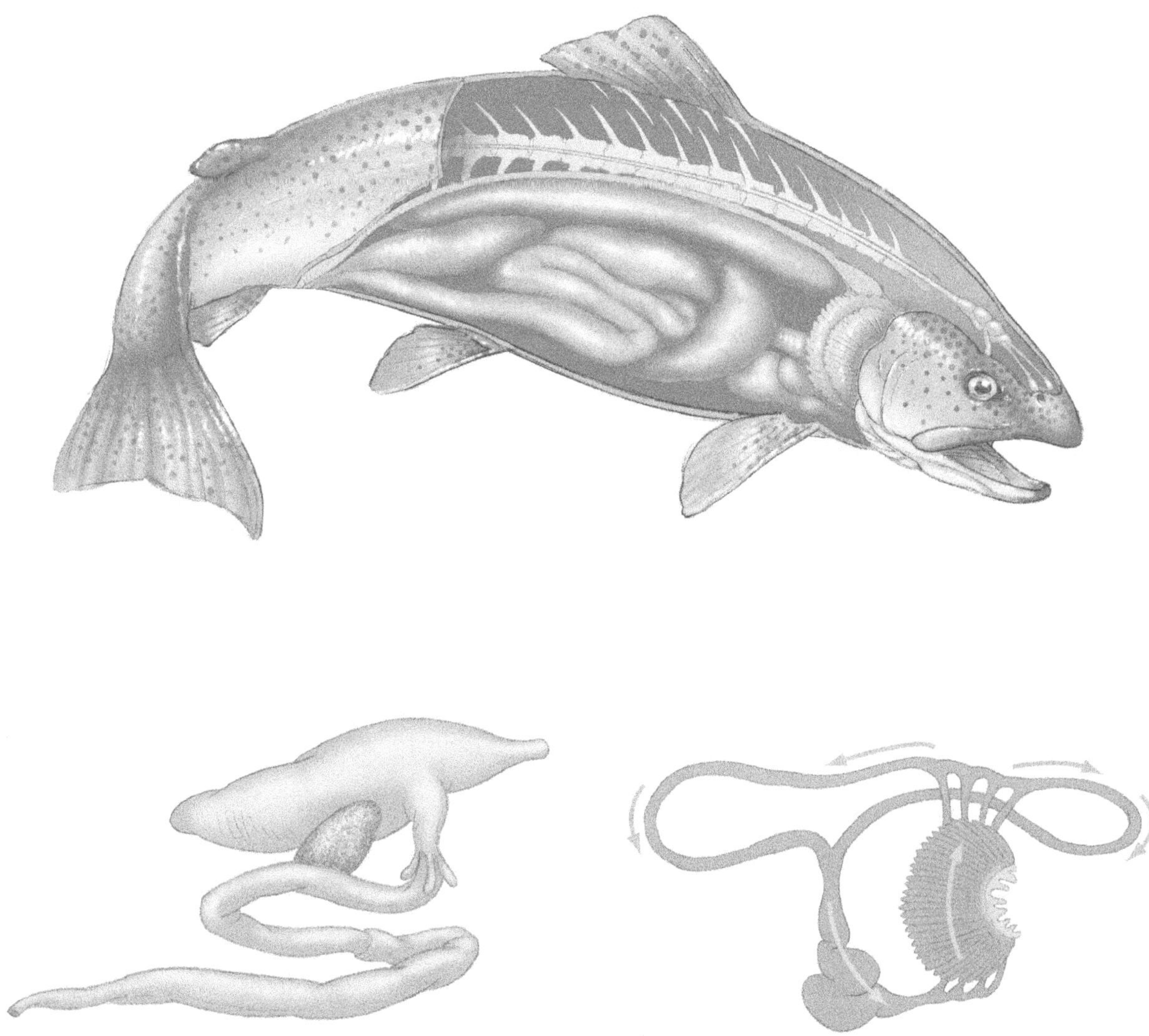

Figure 12.1. Fish.

III. Phylogenetic Placement

Work in your group to draw a phylogeny that includes the lineages from the last lab plus the Myxini, Petromyzontida, Chondrichthyes, Actinopterygii, and Sarcopterygii. Map the following traits on the interior branches of the tree:

- Bone
- Cartilage
- Closed circulation
- Coelom
- Ctenoid scales
- Cycloid scales
- Diphycercal tail
- Dorsal nerve cord
- Endostyle
- Eyes
- Ganoid scales
- Gills
- Heterocercal tail
- Homocercal tail
- Jaws
- Lobe fin
- Lungs
- Nostrils
- Notochord
- Operculum
- Paired appendages
- Pharyngeal slits
- Placoid scales
- Post-anal tail
- Ray fin
- Swim bladder
- Vertebrae

Use the lab report template provided by your Instructor to complete this exercise, save the document as "Lab 12 Phylogeny (last names of group members)," and turn the document in as instructed.

IV. Mini Study

Work in your group to collect behavioral data on shoaling behavior in fish.

- Discuss the information you gained in your independent research on the functions of shoaling behavior with your group members.
- Use this information and the resources provided in lab to form and test a hypothesis that predicts the shoaling behavior of fish.
  - **Hint:** Consider testing the importance of ONE of the following sets of variables:
    - presence or absence of cover vegetation
    - variation in group size
    - variation in individual coloration and patterning
    - variation in the size of individuals
- Enter your data into an Excel worksheet and produce appropriate figures that will allow you to draw conclusions.
- Complete a lab report that contains the following:
  - An introduction including both your null and alternative hypotheses, and a purpose statement.
  - A description of your materials and methods, including statistical methods.
  - A results section discussing your initial findings
  - A copy of your data in table format and any figures used to interpret your data.
  - A conclusion section citing two scholarly articles.
  - A critique of your study: How would you do things differently in the future?

You will likely not have time in lab to complete the analyses and report. Discuss when and where ALL group members will meet to complete the analyses of the data and writing of the report.

- Save the document as "Lab 12 Mini Study (last names of group members)" and email it to your teaching assistant by midnight on Sunday.

V. Examples of the Representative Taxa

Make notes about the various examples of each taxa on display.

LAB
13

Vertebrata II–Tetrapods

This lab focuses on the remaining lineages in the vertebrate clade: the tetrapods. In this lab we will investigate the adaptations that have evolved in these organisms due to the distinct challenges of the terrestrial environment.

Objectives

- Demonstrate knowledge of body types, structures, and associated functions seen in amphibians, reptiles (including birds), and mammals.
- Demonstrate knowledge of and the ability to work in groups to draw a phylogenetic tree with proper placement of key synapomorphic traits for lineages covered in this lab and the previous lab.
- Demonstrate the ability to read and critically assess an article from the primary literature both in writing and orally during group discussions.

Before You Go to Lab

1. Complete Exercise I prior to coming to lab. Completing this section will prepare you for the remaining lab exercises and help prepare you for the pre-lab quiz.
2. Review instructionsfor details on techniques needed to properly dissect your specimen.
3. Attempt Exercise III with your own estimate of the group's phylogeny and associated character placement so that members of your group will be able to discuss differing views and complete the assignment during lab.
4. Read the primary literature article linked to this lab. Download and fill out a Journal Article Review Form to document your critical assessment of this paper. Save the file as "Lab 13 JAR (your last name)." Turn the completed document in as instructed.

Exercises

I. Form and Function

Briefly summarize how the following structures help the animals covered in this lab complete the physiological challenges in the following six categories. How do these traits compare to those in lineages covered in the previous lab?.

Key Terms and Structures

- Bilateral symmetry
- Bone
- Cartilage
- Central nervous system
- Cephalization
- Closed circulation with double circulation
- Coelom
- Complete digestive tract
- Deuterostome
- Dorsal nerve cord
- Eyes
- Kidney (ammonia, urea, uric acid)
- Jaws
- Lungs
- Paired appendages
- Post-anal tail
- Scales/smooth skin
- Triploblastic
- Vertebrae
- Teeth
- Amniotic eggs
- Ear with tympanic membrane
- Internal/external fertilization
- Ectothermic/endothermic
- Wings
- Feathers
- Air sacs
- Hair
- Milk
- Occipital condyles

1. Support and Movement

2. Foraging and Digestion

3. Gas Exchange and Circulation

4. Water Balance and Excretion

5. Development

6. Nervous and Sensory Systems

II. Investigations of Specimens

A. Watch the *Frog Dissection* video for details on techniques needed to properly dissect your specimen and find the following structures. Identify and understand the function of each structure and label figures 13.1 and 13.2.

- Duodenum
- Fat bodies
- Gall bladder
- Large intestine
- Left and right lungs
- Left atrium
- Pancreas
- Right atrium
- Right, left, and median lobes of liver
- Small intestine
- Spleen
- Stomach
- Truncus arteriosis
- Urinary bladder
- Ventral abdominal vein
- Ventricle

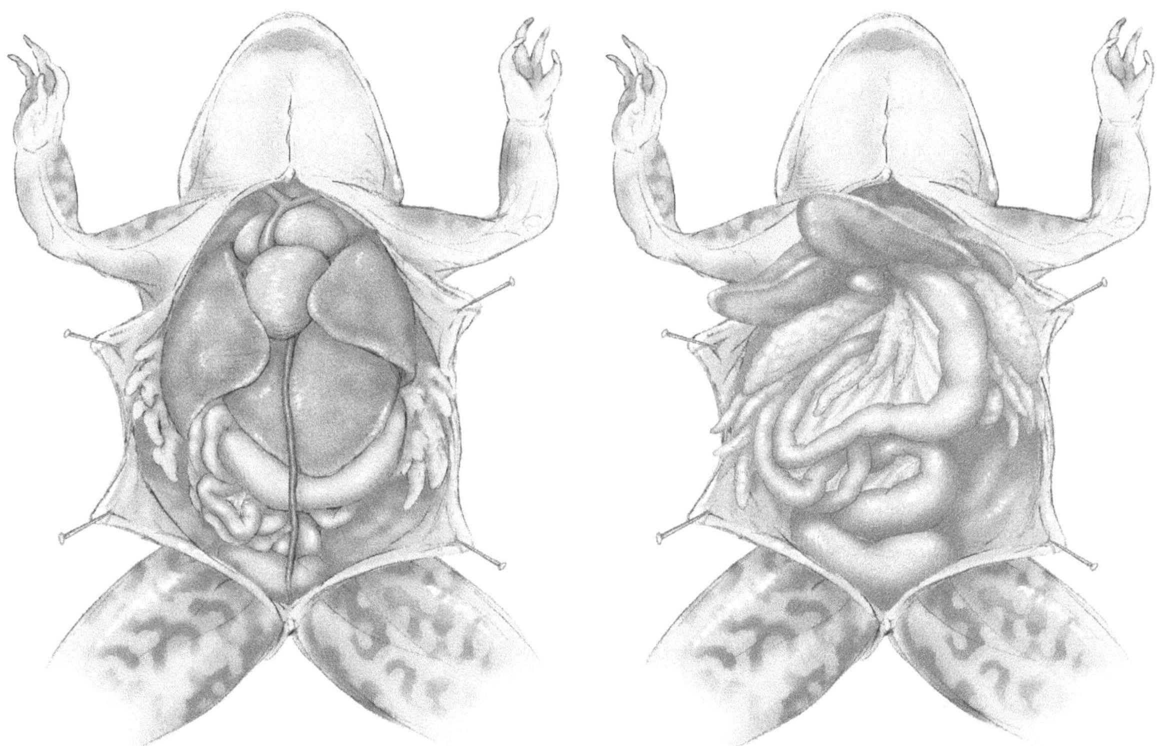

Figure 13.1. Frog.

- Adrenal gland
- Cloaca
- Kidney
- Ovary/testis
- Oviduct/vestigial oviduct
- Posterior vena cava
- Ureter
- Urinary bladder
- Uterus

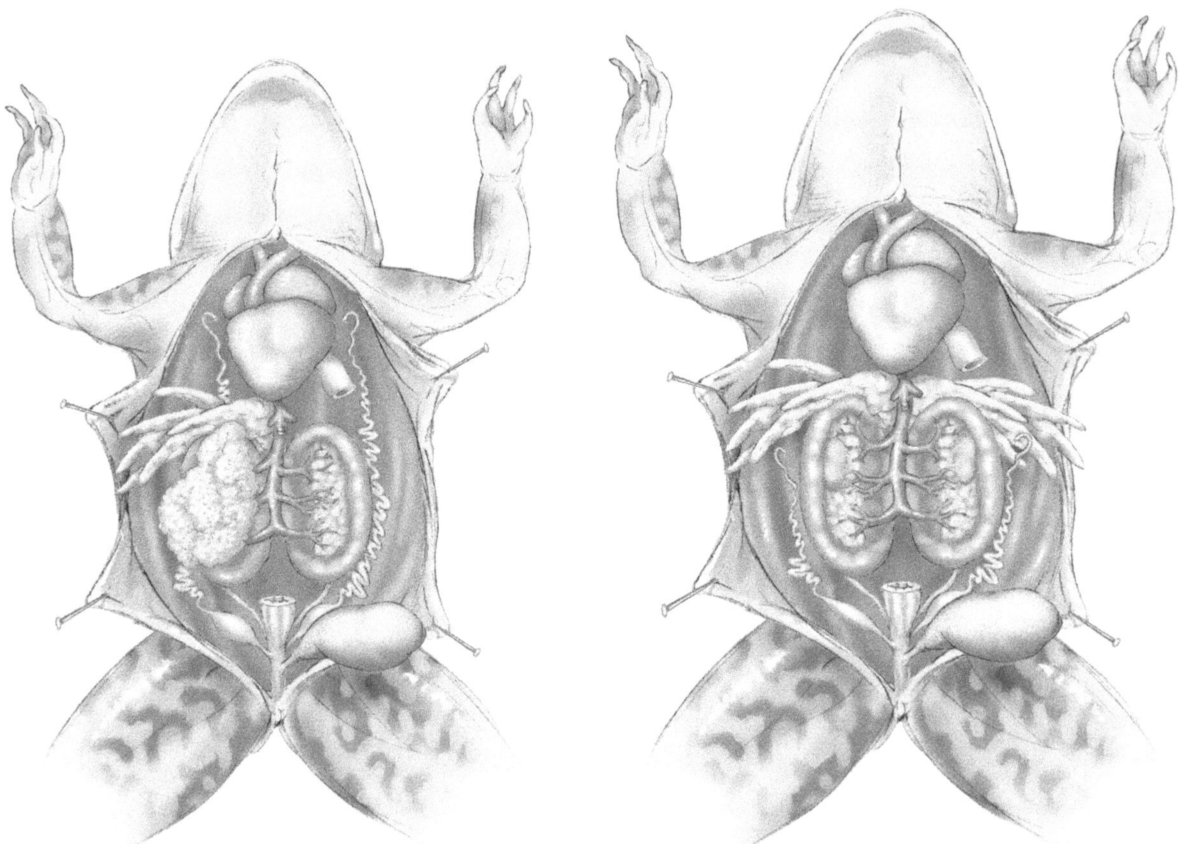

Figure 13.2. Frog.

III. Phylogenetic Placement

Work in your group to draw a phylogeny that includes the lineages from the last lab plus the Amphibia, Mammalia, Lepidosauria, Testudines, Crocodilia, and Aves. Map the following traits on the interior branches of the tree:

- Feathers
- Diapsid, synapsid, and anapsid skulls
- Amniotic egg
- Hair
- Mammary glands
- Paired limbs used for terrestrial locomotion
- Modifications of sensory structures for terrestrial habitats

Use the lab report template provided by your Instructor to complete this exercise, save the document as "Lab 13 Phylogeny (last names of group members)," and turn the document in as instructed.

IV. Journal Article Review

Critically review the article provided by the Instructor and answer the following questions:

1. What was the purpose of the study? What questions were the researcher(s) asking?

2. What methods were utilized?

3. What were the main conclusions?

4. What evidence supported those conclusions?

5. Why were those conclusions important?

V. Examples of the Representative Taxa

Make notes about the various examples of each taxa on display.

APPENDIX

A

Care and Use of Standard Lab Equipment

Before starting any lab, it is important to identify and understand the uses for the safety equipment that may be used in lab. Familiarize yourself with the lab safety equipment listed in Table A.1

Table A.1 Lab Safety Equipment

Eyewash station	Used in the event of accidental exposure to the eyes by a chemical irritant.
Emergency shower	Used in the event of accidental exposure to a chemical irritant on the clothing or for extinguishing clothing fires.
Biohazard container	Used to dispose of any item that has been exposed to body fluids
Fire extinguisher	Used to extinguish fires
Trashcan	Used to dispose of normal refuse.
Broken glass box	Used to dispose of any broken glass.
Scalpel blade remover	Used for the proper disposal of scalpel blades.

The above list may not include all safety equipment in your lab. List other equipment and their functions below.

____

____

____

____

Quality dissection equipment is essential to performing correct and identifiable dissections on specimens. Instruments should be kept clean and sharp at all times. If you suspect that your scalpel or scissors are dull, notify your lab instructor immediately. Identify the instruments used in this lab and understand their functions (Table A.2).

Table A.2 Common Dissection Instruments

Dissecting gloves	Protects your hands from preservatives. May be constructed of either nitrile or latex.
Dissecting pan	Holds the specimen to be dissected.
Scalpel	Used for making incisions into and through the body wall.
Scissors	Used for making incisions into and through the body wall, especially when there is concern about underlying structures.
Blunt probe	Used to inspect and manipulate specimen.
Forceps	Used to loosen organs from tissue, teasing apart tissue and lifting and manipulating part or the entire specimen.

(Continued)

Table A.2 Common Dissection Instruments (*Continued*)

Dissecting needles	Used to loosen organs from tissue and teasing apart tissue.
Dissecting pins	Used to hold specimen to dissecting pan
Ruler	Used for measuring specimens or parts of a specimen.
Pipette	Transfers small volumes of liquids

The above list may not include all dissection instruments in your lab. List other equipment and their functions below.

Understanding the Specimen

APPENDIX B

Biologist use a specific set of terms when describing the anatomy and orientation of a specimen. Review the terms in Tables B.1 and B.2 and identify them on Figure B.1.

Table B.1 Body Planes

Frontal Plane	Plane that bisects a bilaterally symmetrical animal into dorsal and ventral halves.
Sagittal plane	Longitudinal plane passing from head (cranial) to tail (caudal). Bisects a bilaterally symmetrical animal into left and right halves.
Transverse plane	Plane that bisects a bilaterally symmetrical animal into anterior and posterior halves.
Cross-section	A cut made through the body along a transverse plane.
Longitudinal	A cut made lengthwise through the body, parallel to the long axis of the body.

Table B.2 Directional Terms

Oral	Toward the mouth; the opposite of aboral
Aboral	Away from the mouth; the opposite of oral
Anterior	Toward the front of the body; the opposite of posterior
Posterior	Toward the rear of the body; the opposite of anterior
Caudal	Toward the tail end; the opposite of cephalic
Cephalic	Toward the head end; the opposite of caudal
Deep	Away from the surface of the body; the opposite of superficial
Superficial	Toward the surface of the body; the opposite of deep
Distal	Away from the point of attachment, usually in reference to a limb; the opposite of proximal
Proximal	Toward the point of attachment, usually in reference to a limb; the opposite of distal
Dorsal	Toward the back or upper surface; the opposite of ventral
Ventral	Toward the underside or below; the opposite of dorsal
Lateral	Toward the side; the opposite of medial
Medial	Toward the midline; the opposite of lateral

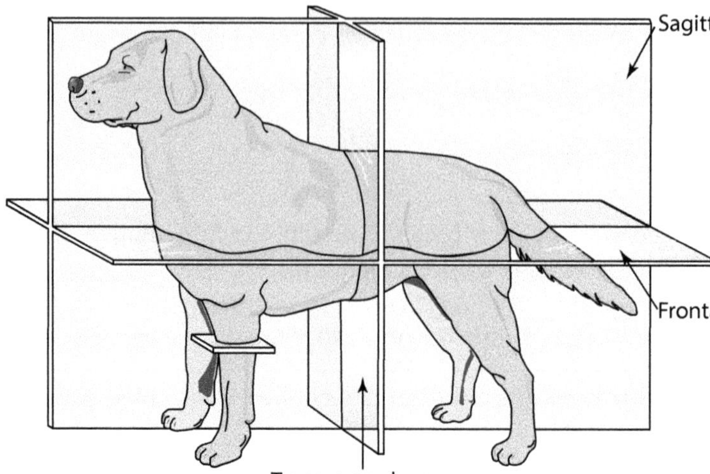

Figure B.1. Body Planes

www.ingramcontent.com/pod-product-compliance
Ingram Content Group UK Ltd.
Pitfield, Milton Keynes, MK11 3LW, UK
UKHW051508290926
15471UKWH00008B/1489

9 781465 283016